Genetics, Ethics and Parenthood

# Genetics, Ethics and Parenthood

EDITED BY

## Karen Lebacqz

*The Pilgrim Press*　　　　　*New York*

Copyright © 1983 The Pilgrim Press
All rights reserved

No part of this publication may be reproduced, stored in a retrieval
system, or transmitted in any form or by any means, electronic,
mechanical, photocopying, recording, or otherwise (brief quotations used in magazines or newspaper reviews excepted), without
the prior permission of the publisher.

Scripture quotations are from the *Revised Standard Version of the
Bible*, copyright 1946, 1952, and © 1971 by the Division of Christian Education, National Council of Churches, and are used by
permission.

**Library of Congress Cataloging in Publication Data**

Lebacqz, Karen, 1945–
Genetics, ethics, and parenthood.

Bibliography: p. 103
1. Medical genetics—Moral and ethical aspects.
2. Human reproduction—Moral and ethical aspects.
3. Parenthood—Decision making.   I. Title.
RB155.L38      1983       174.25       83-2243
ISBN 0-8298-0671-7 (pbk.)

The Pilgrim Press, 132 West 31 Street
New York, N.Y. 10001

# CONTENTS

# ILLUSTRATIONS

# CONTRIBUTORS

*Karen Lebacqz* is Professor of Christian Ethics at Pacific School of Religion in Berkeley, California. She has worked extensively in the field of bioethics for more than ten years, particularly on ethical issues related to research on human subjects, genetics, and mental retardation.

*Richard Ramsey* is Professor of Biology at Rocky Mountain College. He has maintained an ongoing interest in genetics, beginning with his graduate studies at the University of California in Davis. The interface between Christian faith and scientific inquiry has been an essential component of his academic and church involvements.

*Nancy and Michael Rion* are the parents of three children; Carter, their firstborn, has Down's syndrome. Nancy has taught English and writing in high school and college and has served as Director of Christian Education in a UCC church for two years. Michael is Corporate Responsibility Director at Cummins Engine Company in Columbus, Indiana. He is a Ph.D. candidate in religious social ethics at Yale University.

*Roger L. Shinn* is Reinhold Niebuhr Professor of Social Ethics and Counselor to Graduate Students at Union Theological Seminary in New York. He is adjunct Professor at Columbia University and at New York University Graduate School of Business Administration, and Visiting Professor at the Jewish Theological Seminary of America. He is a past president of the American Theological Society and of the Society of Christian Ethics. He has traveled extensively for the World Council of Churches in

Asia, Africa, and Eastern and Western Europe, studying issues of the relations between technology and social ethics. In the United States he chaired the National Council of Churches' Task Force on Human Life and the New Genetics. His most recent book is, *Forced Options: Social Decisions for the 21st Century.*

# Introduction:
# Setting the Stage

Genetics! Science, superstition, prejudice, destiny—the word genetics evokes a range of emotions and reactions. To some, it is reminiscent of programs of sterilization, persecution, and racism. To others, it represents our hope for the future of humankind or for a more manageable present. Many of us grew up in a day when the word was virtually unknown—when there were no "genetic counselors" or screening programs, and little that we could do about "hereditary" disease. Others growing up today will take for granted the existence of genes and of our ability to manipulate them.

Whatever else genetics may connote to you, it represents a new human power that, like all human powers, can enhance or distort the meaning of our lives. And this power presents issues for the church, for troubled families, and for all people concerned about personal and social values.

How shall we handle this power? What does our faith say about it? Is there a "Christian" response to new developments? These are some of the issues to be addressed in this volume.

New genetic technologies offer two distinct kinds of power related to the biology of conception, gestation, and birth. The first power is the power to become parents. Artificial insemination by donor (AID), surrogate motherhood, in vitro fertilization (the "test tube babies" popularized in the media), and the possibilities of embryo transfer expand the options for couples and individuals wanting to become parents but facing some difficulty. They also extend the range of people who are involved in some stage of the bearing and rearing of children. A woman who has no desire to raise a child but wants to experience pregnancy might offer to be a "surrogate mother" for another woman who is unable to bring a child to term. A man who does not want to raise a child might give—or sell—his sperm so that infertile couples can become parents through artificial insemination by donor.

The second power is the power to control the "quality" of children. Technologies such as genetic screening and prenatal diagnosis offer the opportunity to predict or diagnose genetic abnormalities prior to conception or birth. Artificial insemination and in vitro fertilization might also play a part in avoiding genetic anomalies in children. These technologies present options for controlling the types of children who will be conceived or born.

Clearly, as a result of the possibilities they make available, the new genetic technologies raise questions that demand our understanding and response.

First, new options for conception and birth, and the involvement of new persons in the process, challenge traditional notions of parental roles and biological necessity. What is the meaning of parenthood? Who is responsible for child-rearing? What traits are desirable or essential in parents? What will be the long term effect upon family and community life as eggs, sperm, and even embryos are transferred from person to person?

Second, technologies that provide opportunities and challenges for parents to exercise "quality control" over their children raise questions about parental responsibility and the value of children.

Should nascent life be allowed to develop if it has genetic abnormalities? Abortion is often the mechanism for implementing "quality" decisions, and ethical issues abound in evaluating abortion. But the focus here is less upon the morality of abortion and more upon the implicit values of normal humanity that our discussions of genetic technologies and abortion reveal. What expectations do parents have about "normal" children, and are these legitimate? How do we connect our efforts to eliminate genetic anomalies with our response to living persons in our community who bear the burdens of these same anomalies?

New technological capabilities challenge us to develop a Christian response to some very basic issues about what it means to be human and to live in human community. They challenge some deeply held beliefs about parenthood. Christian tradition has long held that the generation of new human life and the act of "making love" go together. This unity appears to be "split asunder" by technologies that separate the physical act of love-making from the biological generation of new life. Christian tradition has also held that all people are of equal worth. Technologies that offer "quality control" of children appear to suggest that "some are more equal than others."

Many people have raised questions about these new technologies: Should they be used? What safeguards need to be provided? When does their development violate Christian faith? Our purpose here is not to analyze the technologies themselves nor the immediate ethical questions raised by their development and use. Rather, we hope to focus on deep questions about the meaning of parenthood and to see how our understanding of parenthood may change with the advent of these new powers.

Nor are these issues remote from our own experience. One family in our group has a child with Down's syndrome (formerly known as Mongolism). They have known the disappointment of long delays in achieving pregnancy, the profoundly conflicting experience of joy and sadness at the birth of a retarded child, the anxiety of prenatal testing in subsequent pregnancies, and the continuing challenge of parenting a family with a retarded child. Another task force family has a natural child and an adopted child, and has experienced in deeply personal ways the ambiguous connection between biology and parental roles. Another member faces the question whether or not to have a child in her late thirties when the risk of abnormality is increasing.

Almost any group of families in a congregation will find itself touched by some issues related to genetics and parenthood. Perhaps a young couple is disappointed—even guilt ridden—at their inability to bear children. Or another family has a retarded member who is protected and withdrawn in order to save the family public embarrassment. Another couple may contemplate abortion when evidence of genetic abnormality is discovered. In these and other ways, most of us are touched by the pain and bewilderment that accompany the problems and opportunities faced by parents as new genetic and reproductive technologies emerge.

Genetic problems are not merely private issues, then; they affect the Christian community. Both parents and would-be parents cry out for and deserve sensitive nurture and care. We are called upon to enact the loving bonds of community in response to their need. In so doing, we not only minister to these individuals, but we also embody the core identity of the community of faith—an identity that enables the community to move beyond itself.

A focus on genetics and parenthood is therefore an integral part of the tradition of a church community strongly committed to social justice. The church speaks most forcefully in the public arena when its distinctive identity in worship, faith, and caring community is vital as well. Personal dilemmas engender public policy when questions such as genetic screening and programs for the disabled are at stake.

Many of us are troubled by new choices facing parents, by new techniques which seem to contradict deeply ingrained notions about birth and parenting. Here the issue is as much one of faith and human understanding as it is one of choice, and again the identity of the Christian community calls for it to respond. The purpose of this book is to provide a process for individuals and churches that might help to elicit that response. In particular, we look for a response grounded in Christian faith and in the ongoing life of the Christian community.

This book is a workbook of exercises and reflections to be used as a resource in a congregational study group. In Part I we present cases, background material, and theological reflections on issues related to expanding the range of people involved in child-bearing. In Part II, we offer a similar set of exercises focused on issues in "quality control." In Part III we suggest a framework for decision making and some concrete ways for churches to get involved. Finally, the Appendix offers some additional resources on genetics, parenthood, and the ethical issues involved in the development and use of new genetic technologies. For groups that may wish to supplement our discussion of the biological base and theological issues, we have provided a Bibliography from which they may draw.

Throughout this book we focus on parents, their choices and dilemmas, and on what these choices and dilemmas mean to the rest of us. The critical ethical issues of particular technologies are enumerated and argued in various essays, some of which are listed in the Bibliography. Our purpose is to help congregations recognize the dilemmas and opportunities facing parents and prospective parents, to clarify and deepen their understanding of parenthood and of new genetic technologies in theological perspective, and to identify and embrace ways of responding to the issues raised. We offer not a single normative proposal—a "right way" of thinking or responding—but a process whereby a congregational group can consider the issues and find its own insights and guidance.

The materials offered here can be followed in the order presented. But particular groups may have an interest in only some of the concerns enumerated here. For example, if a church is struggling with questions about selective abortion for genetic disease, it might find Parts II and III most helpful. Another church group interested in how to respond to "test tube babies" might choose Parts I and III. Groups may also wish to change the order or to utilize only some of the materials presented in each part. Our goal is fulfilled if the materials facilitate and support a useful process for individuals and congregations.

# PART I

## *The Power
to Be a Parent*

# CHAPTER 1

# New Technologies in Everyday Life

Some new technologies sound exotic—and, indeed, some of them are. But some of these technologies are already in use. They affect real people in their everyday decisions about having children. The following case study is not unusual.

## Artificial Insemination by Donor

Sue and Bill, married for six years, have been trying to conceive for several years without success. They learn through testing that Bill is infertile—his sperm count is very low, and the doctors do not think that he would be able to father a child even with "pooling" (collecting

3

and storing sperm so that the total sperm count is higher when it is injected into Sue's vagina).

Having always loved children, and having assumed they would have a family, Sue and Bill are now struggling with a variety of emotions. Bill feels some vague guilt. Even though he knows that his infertility is not his fault, he feels as though he has disappointed Sue. He also knows that infertility has nothing to do with impotency, but somehow his security about himself as a husband and lover has been shaken. For her part, Sue feels angry—sometimes at Bill, sometimes simply at the situation. She also feels guilty because of her anger. And both Sue and Bill are experiencing profound sadness knowing they will never have "their own" child.

Their doctor suggests a possible solution: artificial insemination by donor, called AID (see box). In most cases, an anonymous donor would give semen; the doctor would check for physical health and for characteristics similar to Bill's. It might also be possible for a relative or friend to donate the sperm. In either case, the child would have Sue's genes, but not Bill's. The child would thus be at least partly "their own" genetically.

Although AID has been practiced for more than twenty years, Bill and Sue do not know of any couples who have used it. They wonder whether other friends have been through this and have simply kept it quiet. They wonder if it is something that should be "kept quiet." They also discover that in such cases the legal issues of parenthood and rights of the child to inherit property are not altogether clear in some states, including their own.

When Sue tells her mother, whom she considers a confidante and friend, her mother says the child would be the result of an "adulterous" union, and that she doesn't want any grandchild of hers to be "illegitimate." Bill's best friend Paul offers to be a donor—and then asks whether Bill and Sue would tell the child who his "real" father is. Paul also asserts confidently that any child of his would be

*Artificial Insemination by Donor*

Semen from an anonymous donor can be introduced into the woman's reproductive tract by a syringe at the proper time during her menstrual cycle for conception to occur. The donor is usually paid a small fee. Sometimes the husband's sperm is mixed with the donor's, so that it is possible that the husband's sperm will fertilize the egg. Selection of a donor is usually done by the medical team, who check for physical characteristics and for any health problems. The couple would not know who the donor is, nor would the donor know who received his sperm. Best results are obtained with fresh semen. The semen is collected by masturbation.

a boy; Sue really wants a girl, and wonders how Paul would respond if it were.

Sue and Bill begin to search their own feelings. With Sue as the biological mother, a genetic link is assured. What is important to them for the other half of the genetic heritage? Does it matter that the child does not carry Bill's genes as well as Sue's? Will the child be "illegitimate"? Will Paul be insulted and the friendship cooled if the child is ugly or has a genetic anomaly? What if Paul is the donor, but the child turns out to be a girl? Or if Paul later insists on telling the child that he is the "real" father? Should the child be told in any case? Who *is* the "real" father? And—most important—what might this experience do to their marriage and to their feelings of love and trust for each other?

STUDY QUESTIONS

Before you read the following questions, try "role playing" this case. Act it out. Put yourself in the place of Bill, Sue, Paul, the

doctor, or Sue's mother. What issues are at stake? Then turn to these
study questions in small groups or alone to help clarify and identify
the ethical and theological issues.

1. Can you identify with Bill's and Sue's intense desire to have their own
   children? Have you ever adopted or considered adoption for yourself?
   What does it mean to have your "own" children?
2. Imagine that you have just learned that you are infertile. How does
   that affect your feelings about your identity as a man or woman? How
   does it affect your sense of who you are—your personhood? How does
   it affect your feelings about God?
3. Consider your "own" children, your "own" parents, your aspirations
   or experience as a parent. What is crucial to parenthood?—biological
   continuity?—sexual intimacy between parents?—loving and raising
   the child?—responsibility for a child's welfare? Can you locate what
   you think the *core* of parenthood is?
4. Would you be willing to donate sperm to a friend? If you saw the
   parents making a big mistake in raising the child, would you want to
   step in? What would you do? Do you think the parents owe anything
   to the donor? If so, what?
5. Would you tell your child? Try to imagine your feelings if you were
   the parents. Now imagine you are the child. What information might
   be important to you about your father? Does the child have a "right"
   to know?
6. Does AID violate your understanding of God's purposes for human
   fulfillment, or is it consistent with those purposes? Why? What resources might you bring to bear from a Biblical perspective to help
   with this decision?

## Surrogate Motherhood

Now reconsider this case assuming that Bill is fertile but Sue has
had a hysterectomy. Bill and Sue experience a deep depression for

the loss of the dream so precious to them. Their yearning to give expression to their marriage through children leads their doctor to suggest use of a surrogate mother. This means that another woman would carry the child. However, it might be possible to use both Sue's eggs and Bill's sperm, if in vitro fertilization is used (see box). Otherwise, it would be Bill's sperm and the eggs would be those of the woman who would bring the child to term. Once again, the child would be partly "their own" genetically—this time, it would have Bill's genes and not Sue's.

---

### In Vitro Fertilization

"In vitro" means under glass. In the technique of in vitro fertilization, eggs are surgically removed from a woman's ovaries, and are fertilized in the laboratory. (This is why the children born are called "test tube" babies in the media. It is not a test tube, however, but a petri dish!) If fertilization is successful and an embryo begins to develop, it is then implanted into the woman's uterus by the medical team. It can also be implanted into another woman's uterus; thus, in vitro fertilization can be combined with "surrogate motherhood."

---

### STUDY QUESTIONS

1. Does your ability to identify with Bill's or Sue's desire to have their own child differ when it is Sue who has the problem? How and why?
2. Would your response be different knowing that: (a) Sue's twin sister, who is single, offered to be the surrogate: (b) a good friend who has four children and loves being pregnant but doesn't want to raise more children herself has offered to be the surrogate; (c) their doctor knows a woman who has been a successful surrogate before and is willing to be one again?

3. Imagine you are the surrogate mother. What might you feel about the child after it is born? Would you be willing to carry a child for a friend or relative? Would you be more willing to carry a child for a couple unknown to you? Why? What would you say to someone in your church who wanted to be a surrogate mother?

4. What do you think of the idea that a woman might "rent" her womb for as much as $15,000? Do you think bodies should be rented? Is this any different from paying the sperm donor? Why or why not?

5. Imagine that you have elected to be a surrogate mother. How would you explain your pregnancy to your six and ten year old children? To your parents?

6. What should be done if the surrogate changes her mind after the child is born and does not want to give it up? Whose child is it? What happens if the child is deformed or "not quite right"? What if it is a very ugly baby?

7. How do you feel about using such extreme and expensive measures to obtain children? Is this procedure just for the rich? Should it be covered by medical insurance and available to all? Should Medicaid pay the cost for poor people? Do you think the world is already over-populated? If so, does that affect your judgment of these exotic techniques?

8. How do you think your faith affects your response to these questions? Do you think laboratory reproduction is wrong or dehumanizing? Does your church have a stand on these issues?

9. What/who is a parent?

Now that we have looked at some "real life" possibilities, we turn to some biological background. The range of new technologies is staggering. We offer only a brief glimpse at the ones that seem most important. After that, we will look at some of the theological questions raised above, and will ask about parenthood in scriptural perspective.

# CHAPTER 2

# Biology and Parenthood

There are two ways of being involved biologically in parenthood: first, by donating gametes (sperm or eggs), and second, by being the bearer of the child as it grows in the womb. We normally assume that both roles are played by the couple who will raise the child— it is their gametes that combine to create the embryo, and the woman whose egg is thus fertilized will bear the child to term. Normally, the man and woman would be husband and wife. Conception occurs when they "make love," or have sexual intercourse.

New reproductive technologies are changing these assumptions. The ones who give the gametes will not necessarily bear or raise the child. Men may donate their sperm with no thought of ever seeing the child, much less being a part of its family. Women who have no intention of raising a child might nonetheless be biologically involved either by donating eggs (ova) or by lending or "renting" their wombs for gestation. Conception of the child may not occur during the natural act of intercourse, but through a complicated series of medical and laboratory procedures. Thus, new techniques are "splitting asunder" roles and relationships that were formerly held

9

together. The risks vary considerably from one technique to the next and could affect the physical, emotional, or legal status of the parents or the child.

The inability to bear a child is often very painful. While adoption has always been available, at least within certain limits, it has not always satisfied the need individuals and couples feel to have a child "of their own." For this reason, efforts have gone into finding ways to overcome sterility. A number of techniques are now available that make it possible for some people who could not formerly bear children to be involved in the process of conceiving and bearing a child, or to have a child that seems more "their own." Mass media have publicized some of these techniques—e.g., the "test tube" babies referred to in chapter 1. Much confusion and misunderstanding still occurs, however. We offer here only a brief description of some of the new techniques that are available, and what they involve for couples who choose to use them. (For a discussion of the human reproductive cycle and the role of genes, see chapter 7.)

# Expanding the Range of People
# Playing a Biological Role in Parenthood

## I. INTERVENTIONS COMPENSATING FOR IMPAIRED FERTILITY IN WOMEN

Under normal circumstances during the years between menarche (the beginning of menstruation) and menopause, one ovum (egg) will mature and be released from the ovaries every twenty-eight days. This process is controlled by a complex interplay of hormones. After the ovum is released it must be able to pass unimpeded through the uterine tubes (fallopian tubes) where it may be fertilized by a sperm.

---

### Interventions Which Expand the Range of Persons Who May Play a Biological Role in Parenting

1. Compensating for impaired fertility in women
   a. Stimulation of ovulation by drug therapy
   b. Microsurgical repair of blocked reproductive passages
   c. In vitro fertilization and embryo transplant
   d. Surrogate motherhood
2. Compensating for impaired fertility in men
   a. Pooling of semen to overcome low sperm count
   b. Artificial insemination by donor

---

When the necessary hormones are not produced in the proper amounts at the appropriate times, the sequence is interrupted. Sixty percent of all structural infertility in women is due to such hormone failure or imbalance. The sequence is also interrupted when the uterine tubes are physically blocked as a result of disease or abnormal development; this accounts for the remaining forty percent of structural infertility. For women experiencing either type of infertility, several approaches exist to enhance their capacity to bear children.

*Hormone Therapy.* For women whose infertility is due to hormone failure or imbalance, various hormone preparations can be administered intravenously at critical periods of the menstrual cycle to stimulate the release of ova. This procedure costs approximately $800 for each month of therapy. One of the older hormone preparations still used in some cases today can produce abnormally high numbers of matured and released ova. If all are fertilized, too many embryos result for a safe pregnancy. In such cases, the embryos are usually aborted. Newer hormone preparations make this much less of a risk.

*Microsurgery.* For women whose infertility is due to blocked uterine tubes, surgery to repair the tubes is sometimes a possibility. Since the uterine tubes are rather narrow, special surgical techniques are required to repair them. The techniques are relatively well developed and pose no unusual risks. These operations are successful in about fifty percent of cases. (This means that about twenty percent of all cases of structural infertility in women can be repaired by surgery.) For those whose operations are unsuccessful, in vitro fertilization and embryo transplant offer the only hope for reproductive success.

*In Vitro Fertilization and Embryo Transplant.*    Eggs can be surgically removed through an incision in the woman's navel. Sometimes hormone therapy is used to enhance ovulation prior to the operation. The eggs are maintained in sterile glassware until semen is mixed with them. If conception occurs and the cells begin to divide, the fertilization is considered successful. The resulting embryos are maintained in glassware for a short period of growth before one or more may be removed and inserted into the uterus by way of the vagina.

Success with this procedure has now been achieved in several countries, including the United States. While the numbers are too small for a comprehensive analysis at this time, congenital defects do not seem to be a significant problem. There is, of course, always a danger of congenital malformation when the embryo is manipulated.

For every successful birth there will be many unsuccessful attempts. Since this is a rather expensive procedure, many couples must face the dilemma of whether or not to try again after one or more failures.

To date the procedure has been limited to married women with obstructed uterine tubes. It is not clear yet when other possible applications might become realities.

*Surrogate Motherhood.*    Sometimes infertility is caused not by problems with the germ cells or uterine tubes, but by the environment in which the embryo/fetus must grow. For example, the woman's uterus may be scarred or missing, or the birth canal may be unsuitable for delivery and a caesarian section deemed inadvisable. In such cases, it might be possible for a couple to enlist the aid of another woman to carry the child to term.

Under existing procedures, the surrogate is artificially inseminated with the husband's sperm. Should in vitro fertilization and embryo transplant become more widely available, these procedures might be used to provide the conceptus; in that case, the surrogate would not be genetically related to the child. Reports suggest that fees for surrogate services may range up to $15,000 or more. Since artificial insemination by donor (AID) has been plagued with legal questions, it seems safe to predict that surrogate motherhood will be faced with the same sorts of questions: who is the child's legal mother? Must the surrogate give the child up if she changes her mind? The emotional investment of the surrogate who carries the child is likely to be far greater than that of an anonymous donor of semen; hence, it is likely that more suits over custody and "parenthood" will arise.

## II. INTERVENTIONS COMPENSATING FOR IMPAIRED FERTILITY IN MEN

Infertility in men is much more of a puzzle than infertility in women. Our methods for treating the condition remain woefully inadequate, partly because we do not understand the causes. Low sperm counts and sperm motility seem often to be at fault, but evidence suggests there is far more to infertility in men than these two variables.

At least one psychologist has suggested that men may experience greater anxiety about infertility than about impotency. Therefore, in spite of methods that compensate for the condition, men will probably have some difficulty adjusting to the situation. Physicians working in the area report this to be the case. At any rate, most cases of male infertility are unresponsive to currently available treatment.

*Pooling of Semen to Overcome Low Sperm Count.*   Occasionally, pooling of the semen of a man with low fertility will increase the chances of conception. In this procedure semen is collected and stored for a period of time until a sufficient amount is available. The pooled semen is introduced into the woman's reproductive tract via the vagina at the appropriate stage of the menstrual cycle.

*Artificial Insemination by Donor.*   If pooling of the man's semen does not enhance fertility and the couple still wishes the woman to bear a child, semen may be contributed by an anonymous donor. (The "donor" usually receives a fee for his service.) The attending physician may assist in selecting a donor with desired physical characteristics. The donated semen is introduced into the woman's vagina in the manner described above. Sometimes it is mixed with the husband's semen.

The effectiveness of this procedure is about seventy percent with fresh sperm and forty-eight percent with frozen sperm. (The couple may even purchase the frozen semen of Nobel laureates!) The cost can be as low as forty dollars per insemination. Biological risks are insignificant, but the legal status of a child born as a result of this method varies from state to state and determining who is the legal father may be complicated. Physicians working in the area report that divorce rates may be higher in this group than in the general population.

## SUMMARY

This quick overview of available techniques for expanding the range of people who can be involved in conceiving and bearing a child serves to demonstrate that some of our traditional notions about parenthood will be challenged by new technologies. While we can sometimes work within the boundaries of techniques that seem to pose no threat (surgical repair, drug therapy), sometimes we extend ourselves into new arenas—removing eggs and sperm from one body to insert them into another, fertilizing eggs in the laboratory instead of inside the woman's body, and so on. While there is no easy dividing line between "acceptable" and "unacceptable" techniques, some concerned parents have begun to raise questions about what it means to manipulate eggs, sperm, and even embryos outside their "natural" environment. To what lengths should we be willing to go in order to produce children "of our own"? Are there limits on our use of the powers of technology?

In the next chapter, we turn to some of the underlying theological questions about the nature of parenthood. A look at Scripture helps set the stage for considering what to do about infertility and how to respond to new technologies. We will then explore these issues further through an exercise that probes the limits of our acceptance of new technologies and raises questions about the power we wield. In the last chapter in Part I, we will turn directly to theological questions about the use of that power.

CHAPTER 3

# Parenthood in Scriptural Perspective

One of the promises offered to anxious parents by new advances in genetic technology is the promise of having children "of your own." To think through the importance of this promise from a Christian perspective, we seek biblical guidance. The science of genetics is new, as are the technologies involved. We cannot expect any direct references to such issues in the stories found in the Bible. What we can expect are general themes that suggest what parenthood means and how to understand the genetic link and its importance in the human community.

What is the significance of children? When are they "our own"? How does a faith stance affect our relationship with children? These are the kinds of questions with which we turn to Scripture.

One overarching theme seems to leap from the pages of the Bible: the importance of covenant in Christian life. In both the Old Testament and the New, the Judeo-Christian story is that of a people bonded together in a covenant in response to God's love.

16

Over and over again the Israelites are reminded of their indebtedness to God's liberating activity and of their responsibility for living in accord with that activity: "You shall not wrong a stranger or oppress him, for you were strangers in the land of Egypt. You shall not afflict any widow or orphan." (Exod. 22:21–22). Specific injunctions for what to do in the Old Testament always seem to derive from and reflect the covenant between God and the Israelites.

Similarly, the early Christian community is reminded that it is a community founded on a covenant with God. Indeed, the very term "New Testament" means "new covenant." While the nature of the covenant may change slightly, the central importance of covenant for understanding who we are and how we should live remains.

The covenant requires responsibility of each for all. Christians are to live as "the body of Christ"—one body, with each bringing her or his special gifts. Paul addresses the churches with the greeting "brethren"—a sense that all are as close as family members. The covenant is extended to gentiles as well as Jews, slaves as well as free, male and female alike. Our actions are to reflect our existence "in Christ"—in the covenant.

This fundamental theme of covenant responsibility lays the groundwork for an examination of parenthood in biblical perspective. Since responsibility *to* and *for* the other is a constant theme for all Jews and Christians, we must ask in what ways parenthood brings a special kind of responsibility. Since community and covenant are central to the existence of Jews and Christians in biblical perspective, we must ask whether there is anything special about genetic links in human community.

As we bring these questions to Scripture, we discover that the biblical material yields a twofold approach to the question of parenthood. On the one hand, the role of parents is given great weight, and the bonds among family members are honored and revered. On the other hand, the "family" is extended beyond the bonds of genetics per se and into the larger covenant community. Indeed, some would argue that it includes even the "stranger" or "enemy." Thus, the scriptural material provides a base both for understanding the significance attached to the genetic link in parenthood, and also for moving beyond genetics into a broader concept of "family" and commitment to our "brothers" and "sisters."

In biblical perspective, parenthood is a gift. Children are a sign of God's grace. They are to be treasured as part of the goodness of God's bounty to God's people. Two stories from the Old Testament serve to demonstrate the importance of the bond between parent and child and the meaning of the genetic link.

First is the story of Abraham and Sarah. In their old age, long past expected child-bearing years, YHWH promises Abraham and Sarah a great gift: a child. (Pronounced "yahweh," YHWH is the Hebrew word for God; it is usually translated with capital letters, LORD, in English.) Sarah laughs! But the child is born. Isaac represents their hope for the future, their link with ongoing generations: ". . . through Isaac shall your descendants be named" (Gen. 21:12). Then comes the hardest test of all: God commands Abraham to take his dearly beloved son up on the mountain and to slay him. The enormity of the deed suggests the special link between parent and child. The story has a "happy ending" for Abraham: at the last minute, God stays his hand and saves Isaac.

The second story is that of Job. When Job is tested by Satan, his children are destroyed. At the end of the story, YHWH restores Job's fortunes—including the birth of seven sons and three daughters, all of whom share in Job's inheritance.

These stories suggest several themes for understanding parenthood. First, children are a gift, a blessing. For a couple to remain childless is cause for great sorrow. The fact that it can even be considered a curse suggests that child-bearing and -rearing are understood as ways of responding to God's call. Thus, having children becomes a "vocation," a calling.

Second, a central reason for child-bearing is to ensure the future of the clan, the continuity of generations and the orderly inheritance of property. Hence, the genetic link is an important tie to the family lineage and symbolizes the future of the tribe.

Third, nothing—not even a beloved child—is to come between the faithful and God. The near sacrifice of Isaac serves both to show how important children are and also to remind us that nothing is more important than our basic relationship with God.

These are but two of the many stories from the Old Testament that indicate the importance of the genetic link. The story of Onan is another: Onan sins by refusing to provide children for his brother's clan (Gen. 38:8–10). The laws regarding inheritance also support the importance of genetic bonds. What all these examples suggest is that parenting is a special responsibility and children are a gift given by God.

Nor does the parent-child bond connote only inheritance and the passing on of possessions. The parenting role is brought into play to demonstrate the love and compassion between God and God's faithful—and the pain God feels when the faithful turn away. The relationship between YHWH and the Israelites, or God and the Christian community, is often expressed in terms of a parent-child bond. YHWH loves Israel "like a son" or a daughter. God's people are spoken of as God's children. (cf., Amos 3, where Israel is referred to as one "family"; or Eph. 1:5—"He destined us in love to be his sons through Jesus Christ. . . .") Just as Abraham suffers when asked to sacrifice Isaac, so God as parent suffers at the wrong-doings of God's children. And just as the father of the prodigal son rejoices when his child returns home, so does God rejoice when a "child" of God repents. This imaging of the relationship between human and divine in terms of parenthood is reinforced by references to God as mother as well as father—"like a woman in childbirth, I cry out" (Isa. 42:14).

All of these references tend to suggest the importance of the parent-child bond. If even our relationship to God can be described (perhaps is best described) in terms of parenthood and childhood, then surely the importance of this role cannot be overestimated.

At the same time, the very fact that the relationship between YHWH and Israel, or God and Christians, can be described as a relationship of "parent" to child suggests that the relationship is extended *beyond* the genetic link. And there is evidence in both the Old Testament and the New that this is intended. An extension of "family" beyond genetic links is consonant with God's overall covenant and intentions for humankind.

The story of Ruth is a story of a woman who leaves her own people to go with her mother-in-law and make that woman's people her own. She is adopted into the Israelite community even though she is a Moabite, an enemy. Here, the "family" group is composed of those attached through affection, not blood. Through her faithfulness, Ruth wins a place in society for the widowed Naomi—a task traditionally performed by a blood relative.

Similarly, an important aspect of covenant-fidelity in the Israelite community was providing for those who had lost their genetic links— the widow and the orphan. The laws of the Old Testament are replete with provisions requiring care and nurture for those outside clans and tribes (the widow, the orphan, and the sojourner in the land). Such provisions suggest a covenantal community that goes beyond the genetic lineage of clans or tribes. Family goes beyond genetics, and "parenthood" may need to be understood in broader categories than simply having children "of one's own" in a genetic sense.

In the New Testament, there are indications that Jesus enlarges this tendency. His followers are to dwell as a community that extends the notion of "family" beyond the usual genetic bonds. When he is told that his mother and brothers have come to see him, he retorts: "My mother and my brothers are those who hear the word of God and do it" (Luke 8:21). In like manner, when a woman from the crowd blesses his mother, he retorts: "Blessed rather are those who hear the word of God and keep it" (Luke 11:28). Jesus challenges his followers to be willing to leave even their parents in order to enter the special community of responsible covenant that becomes the body of Christ (cf., Luke 18:29–30). Indeed, John the Baptist makes it clear that genetic lineage saves no one (Matt. 3:9). The community of belief is stronger than the genetic bond. (Our tradition of having "godparents" lies along these lines.)

Following Jesus, Paul also extends family beyond genetic lines. By the very fact that Paul calls his fellow-Christians "brothers" and "sisters," he indicates a broader understanding of family. The church is the "household of faith" (Gal. 6:10). Our kinship with each other is by "adoption" (Gal. 4:5; Rom. 8:23). Indeed, Paul goes so far as to urge his friends not to marry and start families. While this urging is sometimes understood as an indication of Paul's expectation of the imminent return of the risen Lord, it is clear in general that he follows Jesus' lead in pushing the image of the family beyond genetic bonds.

Indeed, contemporary Christian ethicist Stanley Hauerwas argues that the early Christian community was distinctive precisely because it did not depend on genetic links or child-bearing for the future of the community. The fact that singleness was encouraged suggests a community that does not need to have children in order to ensure its survival. This then raises a crucial question: what *is* the purpose of having children? Since all within the covenant are to care for each other, Hauerwas suggests that the purpose of having children in such a community is to validate the special bonds of care-taking that go with parenting. Parenthood becomes a true vocation, as is already intimated in the Old Testament. Children become a sign of a community that can hope and trust sufficiently to permit special covenants to be formed.

Finally, the "end time" is sometimes depicted by New Testament writers as a time when family members turn against each other ("and brother will deliver up brother to death . . . and children will rise against parents. . . ." Mark 13:12). Thus, the coming of the Reign of God (traditionally, the Kingdom) appears to require destruction of normal bonds and assumptions. Whether the breaking down of the genetic bond is seen as a negative event in apocalyptic terms or a positive one in prophetic style, the basic point is clear: in the "kingdom," and in the community challenged to live in Christ, genetic links are not as important as covenant fidelity to that community and to God.

As we review these passages from Scripture, two things become quite clear. First, parenthood is a very important role. If it were not, it could not be used as a metaphor for God's relation to us, or for our relations to each other. Second, and equally important, those who are called into covenant by God are called into a covenant that is not dependent upon genetic bonds for the closeness and responsibility of "family."

Perhaps the best way to encapsulate both tendencies in the scriptural material is to see them as indicating that we are all "children of God." ("Child of God" has special meaning still in some parts of the country, where it refers to those who are mentally retarded or infirm. Such a use is consonant with the biblical perspective.) This term connotes both the importance of "children" and the sense that none of us "belong" to each other as much as we "belong" to God. All people—and hence, all children—are gifts from God to the human community. There is no "ownership" of children, then, but only the response to God's gift. If speaking of children as "our own" is linked to concepts of ownership, we risk misunderstanding the nature of the gift.

Understood this way, we can see both why the role of parent is a special one deserving of respect—"honor thy father and thy mother"—and also why it is not limited to those with whom we have genetic ties. Understood in biblical perspective, parenthood is both a special gift and a general responsibility of each for all. The special bonds of family covenants make sense within the larger bonds of the community covenant with God. The normal way of caring for small, dependent children is through the biological family; thus, biological families are revered and supported. At the same time, all are God's children. Where biological ties are not available (as with the widow and orphan), it is the responsibility of the community to become "family." To bear a child is a great gift, but to adopt a child or care for another's child is equally a part of what it means to belong to the covenant of God's people.

### QUESTIONS FOR DISCUSSION

1. What image of the family seems most consonant with biblical stories? What other stories might you use?
2. What does it mean for parenthood to be a vocation? Is parenthood different in a Christian community than in other communities? In what ways?
3. What do you understand God's basic promise to be?
4. In what ways might your church be a better "family" for those whose biological ties do not give them a family?
5. Does a biblical view seem to support the use of new technologies such as in vitro fertilization or artificial insemination by donor? How does your discussion of this and the other study questions help Sue and Bill with their decision?

# CHAPTER 4

# Testing the Limits

We saw in chapter 2 that many techniques are being developed to help people become parents. As indicated in chapter 1, some of these are already in use in the everyday decisions of Christians and others. But some of the new techniques still seem very "exotic." And they hold out the possibility not only of helping infertile couples to have children, but of permitting all sorts of people to be involved biologically in parenthood.

One such technique is in vitro fertilization.* While its primary use is for couples who have agonized many years over their inability to have children, in vitro fertilization also holds out the prospect of "renting wombs," transferring embryos from person to person, and other interventions that go beyond simply helping couples to have children.

The following exercise is designed to explore some of our feelings about biology and parental roles by looking at some of the possibilities

---

*A description of in vitro fertilization can be found on pp. 7 and 12. The exercise is adapted from a design developed by Carol Severin, San Francisco Theological Seminary, 1979.

that in vitro fertilization might bring. The suggested procedure for using this exercise is this:

1. Complete the questionnaire individually. Save the discussion questions for small group discussion.
2. Share initial reflections and concerns in small groups.
3. Next, turn to the discussion questions (still in small groups).
4. Now, in the whole group, identify key issues for subsequent exploration. The discussion questions suggest the range of issues which might be pursued. The Bibliography will provide further references.

**EXERCISE 1:**

The following list provides a series of possible situations for in vitro fertilization. Each scenario can be ranked unacceptable, neutral, or acceptable, indicating whether you would approve in vitro fertilization in this instance. Take a few minutes to indicate your initial response to each of these scenarios:

1. A man and a woman are married. The woman has blocked tubes, so her eggs cannot reach her uterus. They desire a child "of their own." Having tried surgical repair of the woman's oviducts, they now seek in vitro fertilization.
2. A man and a woman are married. The wife cannot produce fertile ova, but wants to carry a child for the couple. The wife's sister offers to donate her ova for fertilization by the husband's sperm.
3. A man and a woman are married. The wife has blocked tubes, and the husband is infertile. The couple wants to fertilize the wife's ova with the husband's brother's sperm in vitro, and transfer the embryo back to the wife.
4. A man and a woman are married. The wife cannot produce fertile ova, but wants to carry a child for the couple. An egg donated by a woman resembling the wife will be fertilized with the husband's sperm, and the embryo will be transferred back to the wife.
5. A man and a woman are married. Neither one can produce fertile gametes, but they want to have a child. The wife's sister will donate ova and the husband's brother will donate sperm. The embryo will be implanted in the wife's uterus.

6. A man and a woman are married. The wife had a hysterectomy, but they want a child linked to them genetically. The wife's sister offers to carry the child fertilized in vitro. The gametes used will come from the husband and wife, and they will raise the child.

7. A man and a woman are married. The wife was born without a uterus. They want a child of their own. They decide to contract with a surrogate mother. It is legally agreed that the surrogate cannot be made to undergo prenatal diagnosis or abortion.

8. A man and a woman are married. They desire to have a child, but the wife doesn't want to interrupt her career with a pregnancy. The husband's sister offers to carry the child for them.

9. A man and a woman are married. The wife cannot carry a child. They plan to fertilize the wife's sister's ova with sperm from the husband. They will pay a fourth party to carry the child.

10. A man and a woman are married. Both husband and wife have genetic defects that they do not wish to pass on to their children. They wish to use in vitro fertilization with gametes from donors who resemble them, and then have the embryo transferred to the wife.

11. An unmarried heterosexual couple has tried to conceive, but cannot because the woman has blocked tubes. The couple requests in vitro fertilization to have a child "of their own." They plan to share child-rearing responsibilities. They might marry some day.

12. An unmarried heterosexual couple wants a child "of their own," but neither can produce gametes. The woman's sister will donate ova and the man's brother will donate sperm.

13. An unmarried heterosexual couple desires a child of their own with a genetic link, but the woman has had a hysterectomy. Her best friend offers to be a surrogate mother to bear the child.

14. A young woman with a good income desires to be a single parent. She has blocked tubes. She wants to fertilize her ova in vitro with her lover's sperm and carry the embryo to term.

15. An infertile single woman on welfare desires to birth and raise a child. She wishes to fertilize a child in vitro with anonymous donor sperm.

16. A single man wants a child of his own. He purchases ova from an attractive woman to be fertilized by his sperm, and hires another woman for host gestation.

17. Two homosexual women, committed to one another for life, desire to have a child. To share the experience, one will contribute ova, the other her womb. Sperm will be provided by donor.

18. Two homosexual women, committed to each other for life, desire to have a child. Both have diabetes and do not want to risk passing the disease on to their child. Using donor ova and sperm, they hope that one of them will be able to carry the child to term.

19. Two homosexual males desire to raise a child. One man's sister donates ova, the other man donates sperm. Another woman is hired to bear the child.

Stop here until you have shared your initial reflections, then go on to the discussion questions that follow.

**DISCUSSION QUESTIONS FOR EXERCISE 1:**

1. How would you define "parent"? What is essential? How would you define "family"? What is essential? Is marriage essential to these definitions?

2. What responsibilities do each of the following have for child-rearing?:
   • biological parents
   • child's apparent mother and father
   • other family members
   • church community
   • wider community

3. How should prospective parents weigh the relative options of new reproductive interventions for childbirth and adoption? What factors seem most important?

4. How much should parents rely upon natural imperatives and how much upon their ability to choose new possibilities? If you draw a line somewhere between a strict "natural law" approach that permits no interventions into natural processes and a "freedom" approach that permits complete freedom to choose, where—and how—do you draw the line?

5. Try to imagine the consequences for a family of pursuing some of the more "exotic" possibilities for attaining birth. What opportunities for human fulfillment and what stresses and strains might this place on

the family? How can the church best give support to prospective parents facing these stresses and possibilities?

6. How does this discussion relate to your understanding of God's grace and sovereignty? Are parents who seek new reproductive interventions displaying sinful pride and selfishness or do they manifest God's creative intention for human self-realization?

7. If you had friends in situation #4 above, and they asked you what "Christians" think about what they are planning to do, what would you answer?

By "testing the limits," we have raised questions about our use of new technological powers. We turn now to a biblical perspective on technology and power.

# CHAPTER 5

# On Power, Technology, and Ethical Decisions

In chapter 1, we saw that one of the questions raised by new genetic technologies is what it means to have children "of your own." Thus, we turned to the Bible (chapter 3) to look for those underlying themes that might help us to understand the place of parenthood in Christian life and the meaning of the genetic bond between parent and child. While the Bible does not address these questions in the same terms that we do today, it nonetheless provides a "deep structure" for thinking through such concerns.

The same is true as we turn to another central question that arises from the development and use of new technologies: Are there limits to technological power? Are there things that we might do but should not do?

As before, the Bible gives no explicit answer to many ethical questions concerning the development and use of technology. It does not define moral responsibility in driving automobiles or flying airplanes, does not prescribe a just system of taxation or social security, does not tell the best ways to feed a hungry world.

Indeed, the method of the Bible is not to answer all our questions. It is not a rule-book telling us what to do or not to do. It is the witness of a people to God's presence in the world and in their lives. It tells us who we are, and how to learn about who we are. Above all, it tells us who we are in relation to our Creator-Savior and to our neighbors. Learning who we are, we may recognize what we are called to do. With some searching and some praying, we may discover responsible patterns of living in our strange new world.

With this in mind, we turn to three passages of Scripture that address questions of power and technology. While they do not tell us specifically whether it is right to develop or use genetic technologies, they provide a "deep structure" for looking at questions of the use of technological power.

### CREATION AND THE IMAGE OF GOD: GENESIS 1–3

The biblical story of creation (often considered as two stories: Gen. 1:1–2:3 and Gen. 2:4–3:24) expresses the conviction that God's human creatures are unique in respect to the rest of creation. We are created "in the image of God." That means both privilege and responsibility. We are given a "dominion" over other creatures and a duty to till and keep the garden that God has planted.

This story communicates several insights that contribute to human identity as Christians understand it. From the creation stories we affirm that life is a gift to be acknowledged in gratitude, and indeed that the whole created order is a wonderful gift. We also understand human beings to have a derived glory (as the "image of God"), limited but real power over our environment, and responsibility for a garden that needs our tending.

These convictions might be summarized in four basic affirmations. First, the creation is *good*, indeed "very good." Second, good is easily turned to *evil*, especially if people strain to be "like God." This implies, third, that we are *creatures* who must recognize our limits. Finally, however, we also affirm that we are *responsible stewards* who are given some powers to be used for God's purposes.

It has become popular in the ecological movement to blame the Christian doctrine of dominion for human destructiveness in the natural order. This charge is often far too glib. The biblical account of creation does not endorse destructive exploitation of nature, and much of the Bible (e.g., many of the Psalms, the prophets, and Job) expresses awe before nature as a reminder of the power and goodness of the Creator. The story of creation holds up our *creatureliness* as well as our *freedom and power*.

Moreover, our power is accompanied by *responsibility*. This is best seen by comparing human power with the power held by other creatures. Microbes can make the soil fertile or can kill animals and people. Lions can eat lambs and can become food for vultures. Non-living processes also have power: rains and wind can nourish crops and cleanse the atmosphere; they can also destroy human homes. Thus, these other creatures and processes also have power. But they do not *decide* to do whatever they do. Human "dominion" is different from the power of natural processes because it is purposive. It is directed toward human interests and purposes. And because human beings know (in part) what they are doing and decide (in part) what to do, they are morally responsible for their acts, as microbes, beasts, and thunderstorms are not.

## THE PERPLEXITIES OF POWER: GENESIS 11:1–9

Furthermore, human power can be dangerous. The story of Babel illustrates the problematic nature of human power—and specifically of technological power. The people at Babel wanted to build a "tower with its top in the heavens" in order to "make a name" for themselves lest they be "scattered" and (to use modern language) lose their identity. God does not tolerate this display of power. God confuses their languages, so that the project fails. And the penalty is precisely the fate that people sought to overcome by building the tower: they are "scattered . . . over the face of all the earth."

One of the important things about this story is that at Babel the human aim was not vicious in any moralistic sense. This was not a case of Cain killing Abel, of David lusting after Bathsheba and manipulating Uriah, of Ahab seizing the property of Naboth. The building of the tower was not an act of aggression or imperialism. Compared with many things that people do, the construction of the tower might seem to be harmless, or even a creative act.

The wrong at Babel was more like the hubris of the Greeks than like the crimes of Cain and David and Ahab. We sometimes translate hubris as "pride." But it is not pride in the sense of self-confidence; it is more like an ambitious power that destroys equilibrium. (In Greek mythology, as in the story of Babel, the penalty for hubris was often exactly the fate that the hero-victim sought to avoid.) The trouble at Babel was a false notion of security, a mistaken sense of what it is for a people to make a name for themselves.

The story of Babel echoes Gen. 1–3 insofar as in both cases human beings forgot their creatureliness and stretched toward likeness to God. But in Gen. 1–3, power is good though corruptible; Babel occurs in a fallen world where power is immediately suspect.

The same themes appear in modern questions about genetics and eugenics. Sometimes the ethical worry is about the morally evil misuse of power, as in genetic theories elaborated to assert racial superiority and inferiority. Sometimes the warning is that power is inherently dangerous, in that even relatively innocent aims may turn destructive as experimenters intrude too rashly upon awesome processes of nature. Thus, it is not only the *misuse* of power but the very *existence* of power that may be problematic. Power always leads to the temptation to want to build a tower "with its top in the heavens." When we forget our nature as created beings, we violate God's purposes.

### ETHICAL GUIDANCE IN UNKNOWN TERRITORIES: JOHN 14:12–31

What do we do in the face of the ambiguity of power shown to us both in Genesis and in the story of Babel? We may get some guidance from one of the many passages in Scripture which express an openness to a new future that will be different from the past and that will require unprecedented decisions. Such a picture is what we get from Jesus' charge and promise to his disciples in this passage from the Gospel of John.

Jesus, in a kind of "farewell address," tells the disciples to expect to do *greater works than he has done.* Against the background of the warnings of Babel, that is astounding advice. If the disciples had made any such claim, they would have been brash and irreverent. But Jesus here raises their expectations. How do we reconcile the charge to do mighty works with the warnings of Babel about the use of power?

We note that Jesus does not provide a blueprint for what the disciples are to do after he is no longer with them. Instead, he says that God will send, "the Counselor, the Holy Spirit," who will "teach you all things and bring to your remembrance all that I have said to you." In a characteristically biblical way, the future—with all of its unknown possibilities and surprises—is related to, but not bound by, the memories of the past. Christians live their lives between the promise that is made and shown in the life of Jesus, and the fulfillment of the promise in the "kingdom."

Thus, the Bible renounces the attempt to provide a chart for every future and a code for every moral perplexity. It does not tell our generation what to do with new genetic knowledge. It calls on us

to make responsible decisions. And it relates the promise of guidance
to the memory of Jesus.

In this openness to the future, however, one standard remains:
the mighty deeds are to be done "in Jesus' name." This is not totally
unfettered human freedom; rather, the power is given so that works
of *love* can be done. And Jesus says plainly, "if you love me, you
will keep my commandments." Thus, like the creation stories, this
farewell address gives a twofold message of freedom and of responsibility. Power is given; its proper use is expected. This proper use
is use in God's name and for God's purposes, not for human
aggrandizement.

### REFLECTIONS FOR TODAY AND TOMORROW

From these glimpses into a few biblical texts, we can draw some
theological and ethical insights that may help us to understand and
use new human powers.

(1) First, we act within the tension of acknowledging a gift that
we have been given and simultaneously accepting responsibility for
stewardship over the gift. To fail to acknowledge the gift would be
wrong (hubris!); but to fail to exercise stewardship would also be
wrong.

We need to ask how we can simultaneously cherish a gift and take
some responsibility for it. Human beings have learned through medical and surgical techniques to prevent and heal some illnesses, to
enhance some human possibilities. Consider the following series of
steps toward better health: adequate diet, medicines, surgery, and
genetic interventions. Where shall we draw the line between acceptable and unacceptable steps? Open-heart surgery was as far from
the experience of the Hebrew prophets as genetic therapy is from
the experience of most parents today. Neither is a priori desirable or
forbidden. The task is to use new technologies within the framework
of reverence for the gifts we have inherited.

There are no absolute rules for how to do this. But two cautions might be in order. First, as the seriousness of medical interventions in life increases, the reasons for caution and for reckoning with unexpected consequences also increase. Second, the most spectacular achievements are often tiny episodes in a process of biological creativity that has been going on for millions of years. Every exultation in new achievements must therefore be balanced by concern for this ongoing process and recognition of the marvels of small steps. "Bigger" or "newer" is not necessarily better.

(2) Biological science is a form of power. And power is problematic in the double sense that we have already noted. It can be used viciously, as genetic information and misinformation have been used in eugenics movements. And it can be used rashly with no sense of awe before the mystery of creation.

But power is also a gift and it can be used creatively. Simply shunning power is not the answer. Christians have always believed that God acts in history, both through the natural order and through human acts. It is reported that the first public message sent over telegraph lines was "what hath God wrought!" That showed a reverence for divine creativity acting through human inventions. The same God who acts through pre-human and non-human nature also acts through human beings—and this includes geneticists and their scientific discoveries. It is as wrong to refuse to accept power and responsibility as it is to forget that we are not meant to be in control of everything.

Because of the problematic nature of power, any use of power has the potential to be a place where God is acting, or a place where pride or other sins are manifest. Christians must seek to discern the divine and the demonic in the new human powers of genetics as in all other technologies. The use of technology is not evil per se; the question is whether it is used in accord with God's purposes—with love and in service to others.

(3) This brings us to the question whether some things are "un-

natural" and simply should not be done. Christians, like people of many faiths and philosophies, have often wondered at the creativity of nature and at the relationship between humanity and the rest of nature. Sometimes they have taken "nature," in some sense of the word, as a norm for ethics. For instance, Barry Commoner says that one of the laws of ecology is "nature knows best." This leads to the supposition that things that look "unnatural" are wrong and are to be condemned. For example, removing eggs and sperm from the human body and manipulating them in a laboratory might seem to be wrong if judged against the "natural" way of things.

But the true "natural law" that Christians have discerned and spoken of through the centuries is not simply the workings of "nature." It is rather the discernment of *God's purposes* as those might be evidenced in nature. Thus, the way that things *are* is not necessarily the "natural law." The "natural law" has to do with what God's intentions are for those things. For example, nature's way of preserving the species is to produce many offspring so that a few may survive. If we discern in this that God's intention is for the species to survive, we might choose instead to have fewer offspring and to protect their lives. In this way, human interventions in "nature" may actually better foster the "natural law" of adhering to God's purposes.

There *is* a folly in forgetting the intimate ties between humanity and the encompassing natural order, or the limits that "nature" sets on what we can do. The saltwater in our bodies is derived from the saltwater of the sea, and sodium chloride from the two sources is chemically identical. The rhythms of our activity and sleep are related to the rhythms of the solar system. We are not artifacts, nor is the earth an artifact. (The image of Spaceship Earth, though useful for some purposes, conceals the all-important fact that the earth is not a human construction, and that its limits set some limits on us.) Every step in genetic advance has brought warnings against transforming human reproduction into a factory-like process of manufacture and quality controls. The warnings are in order. But, as in the case of other medical advances, they are not an irrevocable prohibition. The task is to discern God's purposes so that we know what limits need to be respected.

(4) This brings us to the age-old question of ends and means. It is foolish to argue sweeping slogans such as "the ends justify the means" or "the ends don't justify the means." *Some* ends justify *some* means. Some means are useless to achieve some ends. Some means defeat the ends they are designed to attain. For instance, the trouble at Babel was not that building towers is always wrong, or that the wrong tools were used to build this one. But the arrogant building of towers is a mistaken way to establish an identity and ensure security. Much depends, then, on a careful examination of the *ends* we seek and whether those ends are sufficient to justify the *means* we would have to use.

In genetic practice, it is important to remember that ensuring ideal or "perfect" children is impossible in any case. No means will achieve that end. The striving after human perfection may also not be a justifiable end. It is therefore not an end that should "justify" anything and everything that we might do. At the same time, means are available for the prevention of some very serious human ailments. Then the question arises, what ends justify what means? Parents might decide, for example, that it is better not to have children than to run a one-in-four chance of inflicting upon them a very grave disease. If they could use AID to avoid this risk, is this a justifiable means to a justifiable end? Is amniocentesis and selective abortion a justifiable means to prevent the birth of an infant who will have a short and painful life? Would it be justifiable to prevent the birth of a female—or male—child? These are the places where questions about means and ends become problematic, and careful arguments must be made.

(5) Ethical responsibility requires attention to every source of truth, secular as truly as sacred, because all truth ultimately is of God. Responsible behavior in areas of genetics depends on the most ac-

curate scientific knowledge available. But that does not mean that
science becomes the ethical arbiter for human affairs. (Most scientists
strenuously reject that role.) Respect for science without any idolatry
of science is the appropriate Christian stance.

An international, interdisciplinary study group on issues connected
with genetics, assembled by the World Council of Churches, put
the issue this way:

> Church [people] cannot expect precedents from the past to provide answers
> to questions never asked in the past. On the other hand, new scientific
> advances do not determine what are worthy human goals. Ethical de-
> cisions in uncharted areas require that scientific capabilities be understood
> and used by persons and communities sensitive to their own deepest
> convictions about human nature and destiny. There is no sound ethical
> judgment in these matters independent of scientific knowledge, but sci-
> ence does not itself prescribe the good (Charles Birch and Paul Abrecht,
> eds., *Genetics and the Quality of Life* [Elmsford, NY: Pergamon Press,
> 1975], p. 203).

(6) There is no riskless life. Reverence for the gift of human life
means a concern for the ethics of risk. Some risks, accepted pur-
posefully for good ends, are admirable. But there is no nobility in
risks that recklessly jeopardize people and values, above all in risks
imposed on others. The traditional criterion of "informed consent"
protects people in medicine and in medical research. But there is
no way to get informed consent from people not yet born. Thus,
other means of protection are needed for those who might be put at
risk in genetic research.

The customary risk-benefit (or cost-benefit) analysis is also inadequate for genetic experimentation. In human affairs, what matters
is not simply the aggregate of costs and benefits; their distribution
also counts. We must always ask *whose* costs and *whose* benefits are
involved.

An example makes the point. Experimenters have subjected plant
seeds to radioactivity in order to increase mutations. Most mutations

are harmful, but the cost is slight. If a very few are successful, the gain outweighs the loss. Such a procedure applied to human beings would raise cries of justified outrage. Human dignity cannot be reduced to calculations of gain and loss that are allowable for other objects. Inferior plants produced by genetic experimentation can be composted. Inferior human beings produced by a like process are a different story.

## QUESTIONS FOR DISCUSSION

1. How can Christians today relate two biblical insights: the frailty of human beings and their "dominion" over nature? What difference do these insights make for decisions about genetics?
2. When the first "test-tube" baby was born, nobody said "what hath God wrought!" Why not? Is the reason an accident, a change in cultural habits, or something in the nature of the new achievement?
3. What would you consider some modern equivalents of the tower of Babel?
4. How do you assess new technologies such as AID or in vitro fertilization in the light of these scriptural passages about power? Do these new technologies seem "unnatural"? Why or why not?
5. What rules or guidelines might you have now for making decisions about the development and use of new genetic technologies? What sorts of things would make it right or wrong for Sue and Bill to use artificial insemination, for instance?

# PART II

## Parenthood and the "Quality" of Children

# CHAPTER 6

# "Quality Control"

In the previous section of this book we looked primarily at interventions that expand the range of people who can have a role in childbearing. Now we turn to another set of interventions that offer a different kind of power: the power to control the "quality" of our children. These are interventions into the characteristics of children who will be born.

## Case Study: Spina Bifida

Jim and Mary (ages thirty and twenty-nine, respectively) have been married five years and have a healthy three year old boy. They have been eager for a second child to complete their family. A month after Mary conceived, her doctor asked her to repeat a routine blood test taken the week before. Her blood sample had shown an unusually high amount of a substance known as AFP (alphafetoprotein). A high level of AFP might indicate a "neural tube defect" in the fetus.

---

### Neural Tube Defects

There are two major kinds of neural tube defects. (1) In anencephaly, the child has very little brain tissue and usually dies immediately after birth. (2) With spina bifida (also called meningomyelocele or "open spine"), an opening in the spinal column exposes the neural tube. While the opening can be repaired, there is usually paralysis from the waist down and there is some risk of mental retardation. The child will have multiple operations and interventions during its life.

---

When the second blood test also showed a high level of AFP, Jim and Mary got quite concerned. In a discussion with their physician, they learned that the screening process for neural tube defects is a complicated one: out of one thousand women tested initially, fifty will have a high AFP—but most of these are "false positives," in which the child has no defect even though the AFP level has registered high in the test. In order to screen out these false positives, there is a complicated series of additional tests. Those fifty women are then given a second blood test, as was Mary. Perhaps twenty will be cleared; Mary is in the remaining thirty (out of the original one thousand) who still have high levels of AFP. There is still a chance that her child does not have a defect, however. She may now have an "ultrasound" examination, using high frequency sound waves to get a "picture" of the fetus in the womb. The test is not invasive and is considered safe. It will usually clear about half of the thirty women.

If Mary is still among the fifteen women at risk for a defect in the fetus, she and Jim may then choose to have amniocentesis. An amniocentesis gives a more accurate measurement of the AFP level, and most of the fifteen women will be cleared. Of the original one thousand women tested, only one or two will ultimately be found to have a child with spina bifida or anencephaly. Later ultrasound and X-ray tests may identify anencephaly but are not guaranteed to show spina bifida.

In short, Mary's chances of bearing an affected child are only one or two out of one thousand. Still, in order to know for sure whether her child has a defect, she must go through a lengthy and complicated series of tests. The statistics leave Jim and Mary in turmoil. The possibility of an actual diagnosis before birth is attractive. The probability of a healthy child is good, and once the process of testing is over, Jim and Mary could relax and enjoy the pregnancy.

On the other hand, they are already very anxious. While their ultimate chances for a defective child may be small, the initial positive tests have given them some fear. They wonder whether they should abort this child now, before they get too attached to it and before they go through a lengthy and difficult process of screening. Because an amniocentesis cannot usually be done before the thirteenth week of pregnancy, they might otherwise have to consider abortion after the time when Mary has felt the child move and heard the heartbeat. Since they love children and dearly want another one, they don't know whether they could abort a baby after they begin to develop real attachment to it.

Indeed, they have never considered abortion at all before this, and they are not sure whether they could do it. What would they tell their son—and the second child they hope to have someday—if they abort a child because it isn't "perfect"? Do they have the strength to live with an abortion? One study of thirteen families who chose abortion after amniocentesis showed that twelve of the women and ten of the men experienced depression following the abortion; four of the couples (almost one-third) separated. Whatever they do, Jim and Mary seem to be facing a stressful situation.

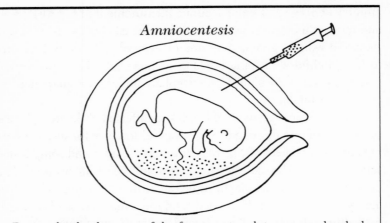

*Amniocentesis*

During the development of the fetus, waste substances are sloughed off and living cells are deposited in the fluid of the amniotic cavity. A trained physician using either his/her own sense of touch or the analytic power of ultrasonic scanning (vague but informative "pictures" taken with ultrasonic waves rather than the more harmful X-rays) can locate the position of the fetus, the placenta, and the amniotic cavity. A narrow, hollow needle is then inserted into the cavity and approximately four teaspoons of fluid are collected. The fluid is sent to a specialized laboratory where the few cells that are present can be encouraged to multiply to the point that they can be more easily located and their chromosomes counted or metabolism analyzed. Occasionally the fluid itself can be examined for telltale chemicals. The procedure which permits the counting of chromosomes, called karyotyping, will cost approximately $400 and takes about four weeks to complete. Collection of fluid can be done effectively only after the thirteenth week of pregnancy. This means that should the tests turn out to be positive, the parents may be faced with an abortion decision well toward the end of the fourth month. Amniocentesis is relatively safe, though some evidence indicates a slight increase in the incidence of miscarriage. Inaccurate diagnosis is infrequent (four out of one thousand); however, its occurrence could have serious consequences. Late abortion carries some risk, though it can be performed "routinely" until about the twentieth week of gestation.

## Screening for Neural Tube Defects

The detection of spina bifida and anencephaly is based upon the fact that after the first month of fetal development a unique protein called alphafetoprotein (AFP) will seep into the amniotic fluid when there is a neural tube defect. This protein may in turn be picked up by the maternal circulation. Hence, starting with a simple blood test, a screening procedure for neural tube defects can be instituted as follows:

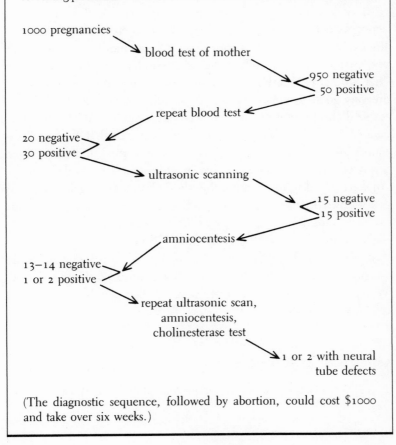

1000 pregnancies

blood test of mother

950 negative
50 positive

repeat blood test

20 negative
30 positive

ultrasonic scanning

15 negative
15 positive

amniocentesis

13–14 negative
1 or 2 positive

repeat ultrasonic scan,
amniocentesis,
cholinesterase test

1 or 2 with neural
tube defects

(The diagnostic sequence, followed by abortion, could cost $1000 and take over six weeks.)

And what if they choose to reject abortion as an option for them? Is their marriage strong enough to withstand the added demands and pressures of a handicapped child? Can they afford the additional medical expenses and services for their child? Can they bear to watch the child be ridiculed and laughed at by other children? Will their relatives and friends accept their child—and them? What will happen to their freedom, their hopes for travel, their plans for retirement, their life-style? Would the child ever wish s/he had never been born?

Mary and Jim are full of conflicting feelings. As they share their concerns with those close to them, they receive varied responses. Jim's mother suggests they abort now to avoid months of uncertainty and anxiety. Mary's mother cries and wonders why God did this to "her little girl." Mary's best friend, Louise, reminds them that they are very good parents—indeed, unusually good—and that a handicapped child might be God's special gift for them. Aunt Susan rails against these "new fangled tests" and proposes that they "just let nature take its course." Friend Sam mentions the hundreds of thousands of dollars "wasted" on his institutionalized retarded cousin. Another friend reminds them that their chances of a defect are only one or two in one thousand, and suggests that perhaps they are "making mountains out of molehills" and urges them to live "one day at a time" and handle crises only *if* they arise.

In the face of all this conflicting advice, Jim and Mary must consider the options and they must make a most difficult decision. Whatever they decide, it is a decision that they will live with for the rest of their lives.

QUESTIONS FOR DISCUSSION

1. Try to imagine yourself in Jim's or Mary's dilemma. What would you do? Why?
   (a) Refuse more screening and "let nature take its course," as Aunt Susan suggests?

(b) Take no chances and abort now, as Jim's mother thinks would be best?

(c) Have an amniocentesis and abort if the results are positive, indicating a defect?

(d) Go through the screening just to know for sure, but not abort even if the tests are "positive"?

2. Has anyone in your church faced this dilemma—or one like it? How might you support Jim and Mary if this happened in your congregation?

3. If Jim and Mary choose to abort the fetus without further testing, the chances are great that they would be aborting a normal fetus. How does this affect your decision?

4. Suppose you had a younger brother with spina bifida. He must be carried or helped with a walker or wheel chair. He has no control over his bladder, so he has to wear a bag; you have to help him with his toilet needs. Your family never goes on vacation because he is always in the hospital or having trouble with his circulatory system. How would you feel toward your parents? . . . your brother? If you were the parents, what would you say to the older child?

5. If you chose abortion above, how would you explain your decision to a paraplegic? If you chose against abortion, what would you say to Jim and Mary?

6. What does your faith say about taking risks? How might being a Christian influence your decision in a case such as this? Where would you go for answers to these difficult questions?

# CHAPTER 7

# Human Reproduction and Genetics

Spina bifida and its related neural tube closure disorders are examples of disorders which are determined to one degree or another by the presence of abnormal genetic combinations in the cells of the person. In spina bifida, the genetic determinants are not clearly understood, but in some other disorders a great deal is known. In this chapter, we will look briefly at human reproduction and the role of genes. We will then illustrate basic types of genetic disorders. Finally, we will say a word about the possibilities for "quality control" of children through genetic counseling and screening.

## The Reproductive Process and the Role of Genes

In humans, conception of new life occurs when the male's sperm fertilizes the female's egg (ovum) somewhere in the fallopian tubes which connect the ovaries to the uterus. In that process, a genetically unique entity, the zygote, is produced. It has the potential for growing into an adult human being. Some growth will occur as the zygote migrates down the tubes into the uterus. As soon as growth (cell division) does take place, this cluster of cells can be called the embryo. It takes about four days for the embryo to travel down the fallopian tubes to the uterus. It must then implant itself into the uterine wall for further growth to occur. Once implanted, the embryo will eventually develop a placental connection to the uterus. This more mature embryo is called the fetus.

There are a number of other important steps in the maturation of the fetus. Spontaneous abortion or miscarriage often occurs quite early in pregnancy—even before the woman is aware that she is pregnant. Twinning can occur until about the third week, resulting in two or more children with the same genetic information. Brain waves are measurable by about the sixth week. Movement can often be felt shortly after the end of the first trimester (the twelfth week). The fetus "looks" human and has human organs and body shape (morphology) by the fourteenth week. With current technology, the fetus is likely to be "viable" (able to live outside the woman's body) by the twenty-fourth week of gestation. All of these stages have been considered important times in the development of a new human being.

Genes play an important role in this development. The characteristics of the infant are determined in large part by the genes contributed by each parent in the sperm and egg.

Genes are arranged like beads on a chain in structures called chromosomes. Humans normally have twenty-three pairs of chromosomes—twenty-two called "autosomal" and one pair of "sex" chromosomes. Each chromosome contains many genes. In the female, both sex chromosomes are the same shape and size; they are called "X" chromosomes. The male, however, usually has one "X" chromosome and one much smaller chromosome called a "Y" chromosome. The "Y" chromosome carries very few genes compared to the larger "X" chromosome. This is why a defective gene on the "X" chromosome will often result in a genetic disease in the man, since there is no parallel gene on the "Y" chromosome to "mask" the effects of the defective gene. Such disorders are called "sex-linked" (or "X-linked") genetic disorders. For example, hemophilia is a sex-linked disorder: if a man gets the gene for hemophilia from his mother, he will have the disease, since he gets no gene from his father that can "mask" or prevent the expression of the defective gene.

Many human traits, such as intelligence, are influenced by the interplay of numerous genes. Others are determined by a pair of genes. The two genes for a particular trait may be identical in information content, or they may be different. If they are identical, then the child will have the trait that those genes represent—e.g., blue eyes. But what happens if the genes are not identical in content—e.g., if one is for blue eyes and the other is for brown eyes?

In this case, what matters is whether the genes are equal in their ability to influence a trait, or whether one of them is "stronger" than the other (that is, more influential). If one is stronger, it is said to be "dominant"; the other is "recessive." In the case of brown and blue eyes, for example, the gene for brown eyes is dominant. If a child inherits a gene for brown eyes from its mother and a gene for blue eyes from its father, it will have brown eyes. But it will also have a "recessive" gene for blue eyes, and so, when the child grows up it could pass on this gene to its children. This is why brown-eyed people can have blue-eyed children. (See picture on page 54.)

All of us normally have many of these "recessive" genes without showing any outward sign of their presence. Only when we inherit two recessive genes for the same trait (and there is no "dominant" gene to mask the effects of the recessive gene) do we show the trait. Thus, in order to have blue eyes, we must get a gene for blue eyes from each parent. Some scientists think that many potentially harmful genes are carried by normal people. They are "recessive" genes that are normally "masked" by dominant genes. We are healthy and normal, because we have the dominant gene. But we can pass on to our children the harmful recessive gene, and if the child gets two of those genes, s/he will have a genetic disorder.

## Genetic Disorders

This brings us to the question of genetic disease. The severity and incidence of genetic disorders can vary significantly. Myopia (near-sightedness), allergies, and abnormally developed wisdom teeth (that have to be removed) are quite common genetically influenced disorders, and they are not usually considered severe. Indeed, we do not usually think of them as "genetic diseases" at all.

At the other end of the spectrum are those disorders that are so profound that they prevent the full development of the embryo and result in spontaneous abortion or miscarriage. Many of these are not even noticed by the potential mother, because they occur so early in pregnancy.

It is the group of disorders that fall in between these extremes of mild and severe that are most important for our consideration, since it is this group that is the subject of most genetic counseling, prenatal testing, and other screening efforts. Table 1 gives the relative prevalence and classification of some of these disorders.

As you can see from the chart on page 55-56, genetic disorders can be grouped into three major categories. (1) Chromosomal disorders are those in which the anomalies relate to the presence or absence of blocks of chromosome material—for example, the "extra" #21 (or #22) chromosome that is present in cases of Down's syndrome. (2) Monogenic (Mendelian) disorders are those disorders strongly tied

## Patterns of Inheritance: Eye Color

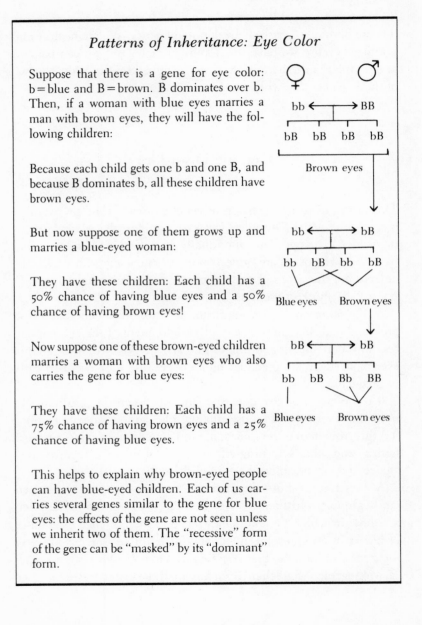

Suppose that there is a gene for eye color: b = blue and B = brown. B dominates over b. Then, if a woman with blue eyes marries a man with brown eyes, they will have the following children:

Because each child gets one b and one B, and because B dominates b, all these children have brown eyes.

But now suppose one of them grows up and marries a blue-eyed woman:

They have these children: Each child has a 50% chance of having blue eyes and a 50% chance of having brown eyes!

Now suppose one of these brown-eyed children marries a woman with brown eyes who also carries the gene for blue eyes:

They have these children: Each child has a 75% chance of having brown eyes and a 25% chance of having blue eyes.

This helps to explain why brown-eyed people can have blue-eyed children. Each of us carries several genes similar to the gene for blue eyes: the effects of the gene are not seen unless we inherit two of them. The "recessive" form of the gene can be "masked" by its "dominant" form.

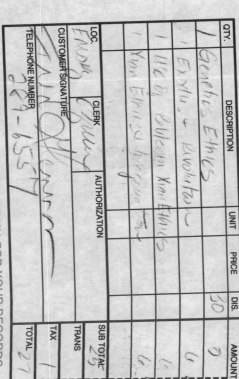

805045320

REV FREDERICK E GLENNON

COKESBURY-EMORY
ATLANTA GA
120386/262900
410430026Y

021192

Cokesbury

| QTY. | DESCRIPTION | UNIT | PRICE | DIS. | AMOUNT |
|------|-------------|------|-------|------|--------|
| 1 | Genetic Ethics | | 30 | 0 | |
| 1 | Exodus & Revolution | | | | |
| 1 | Web Release Xtn Ethics | | | | |
| 1 | Xtn Ethics program | | | | |

| LOC. Emory | CLERK | AUTHORIZATION |
|---|---|---|

CUSTOMER SIGNATURE

TELEPHONE NUMBER
284-6559

| SUB TOTAL | TRANS | TAX | TOTAL |
|---|---|---|---|
| 25 | | 1 | 27 |

CUSTOMER INVOICE—RETAIN FOR YOUR RECORDS

*Estimated frequency of genetic disorders in the newborn population. (Adapted from Epstein & Globus, 1977. "Prenatal diagnosis of genetic diseases." American Scientist 65:703–711.)*

| Disorder | Cases per 100,000 newborn |
|---|---|
| Chromosomal Disorders (disorders involving gross chromosomal abnormalities) | |
|     Trisomy 21 (Down's Syndrome) | 130 |
|     Others (none as prevalent as Down's) | 420 |
|     Total chromosomal disorders | 550 |
| Single Gene Disorders | |
|   Autosomal (not sex-linked) recessives | |
|     Cystic fibrosis | 50 |
|     Albinism | 10 |
|     Phenylketonuria (PKU) | 10 |
|     Tay-Sachs disease | 1 |
|     Others | 179 |
|     Total autosomal recessives | 250 |
|   Sex-linked (on X chromosome) recessives | |
|     Duchenne muscular dystrophy | 20 |
|     Hemophilias | 10 |
|     Others | 20 |
|     Total sex-linked recessives | 50 |
|   Autosomal dominants | |
|     Huntington's chorea | 10 |
|     Blindness (several types) | 10 |

| Disorder | Cases per 100,000 newborn |
|---|---|
| Deafness (several types) | 10 |
| Others | 30 |
| Total autosomal dominants | 60 |
| Multifactorial Disorders (associated with complex combinations of genes and environment) | |
|     Spina bifida and anencephaly | 450 |
|     Congenital heart defects | 400 |
|     Club feet | 300 |
|     Diabetes, allergies, malignancy | ? |
|     Others | 500 |
|     Total multifactorial | 1650 |

Total Frequency of Genetic Disorders = 2%–4%

to the influence of one gene pair. Here, it is the presence of single genes that may make the difference. (3) Multifactorial disorders are disorders for which there is evidence of significant genetic and environmental influence. No single gene pair or block of chromosomal material can be located as the key influence, and the "expression" of the disease appears to depend on a combination of many factors. Since the possibilities for testing and for intervention differ depending on the type of disorder, we will look briefly at some of the unique aspects of these three subgroups.

CHROMOSOMAL DISORDERS

One of the most common best known genetic disorders is Down's syndrome (formerly called Mongolism). Down's syndrome is a chromosomal disorder; it results from the presence of "extra" chromosome material. When a map of the chromosomes is drawn (called "karyotyping"), it is seen that the person with Down's syndrome has an extra #21 or #22 chromosome. (For this reason, the syndrome is also called "trisomy 21," meaning that there are three "somatic" chromosomes in the #21 range.)

Most chromosomal disorders are believed to result from cellular accidents during the maturation of the ova. Some recent evidence also suggests that the production of abnormal sperm may play some role. Whatever the case, these disorders are characterized by abnormal numbers of chromosomes or by abnormal chromosome structure.

Some of these disorders are relatively minor. In Klinefelter's syndrome, the male child has an extra "X" chromosome. Body stature may be affected, but otherwise there are no significant abnormalities clearly associated with the syndrome. In other cases, however, the child is severely affected: a child with "trisomy 13" (an extra chromosome in the #13 range) will usually die shortly after birth. Down's syndrome falls somewhere in between these two extremes. It is characterized by mental retardation and increased susceptibility to a variety of diseases. Some children with Down's syndrome are severely retarded, some are only mildly retarded.

## MONOGENIC (MENDELIAN) DISORDERS

Perhaps the largest number of different disorders falls into the category of monogenic disorders. The inheritance of monogenic disorders may follow several different patterns, resulting in different possibilities for diagnosis and prevention.

"X-linked" disorders such as hemophilia seem to be "carried" on the sex chromosomes, and they afflict primarily males. A girl is not likely to have the disease, though she may be a "carrier" of it and pass on an affected gene to her children. But a boy has a fifty-fifty chance of having the disease. See page 59 for an explanation of the inheritance of sex-linked disorders.

Other monogenic disorders are called "recessive." Here, the child must get one abnormal gene from each parent in order to be afflicted. The box on page 60 illustrates the inheritance of the recessive disorder called cystic fibrosis. Cystic fibrosis affects mainly Caucasians (whites); it is characterized by fluid in the lungs. Until recently, children with cystic fibrosis usually died early in life. Now some survive to adulthood. They face the likelihood of dying by asphyxiation, as their lungs fill with fluid.

Finally, some monogenic disorders (e.g., certain forms of blindness and deafness) are called "dominant" disorders. In this case, it takes the presence of only one gene to "cause" the disorder, since the affected gene is sufficiently strong ("dominant") to overcome any "masking" effects of its normal counterpart. A parent who has such a disorder therefore has at least a fifty-fifty chance of passing the disorder on to any child, no matter how normal the spouse may be. The picture on page 61 shows how dominant disorders are inherited.

MULTIFACTORIAL DISORDERS

Some of the most problematic genetic disorders are those we call "multifactorial." As indicated above, they are not related to the presence of a single gene or gene pair. It is difficult to say exactly how a child comes to have one of these diseases, therefore. Some of them are quite severe. For example, the "neural tube" defects such as spina bifida and anencephaly can be complicated with multiple problems. These disorders result from the failure of the embryonic nervous system to seal properly. Portions of the central nervous system (the spinal column) may be exposed and abnormal fluid pressures may result in seriously abnormal brain development. As noted above, the child with anencephaly usually dies shortly after birth. However, the person afflicted with spina bifida may live a long

*Illustration of the inheritance of hemophilia, a disorder caused by a recessive gene carried on the "X" chromosome (sex-linked inheritance), when both parents are normal as far as blood clotting is concerned.*

Female
cell
2 "X"
Chromosomes

Male
cell
1 "X"
1 "Y"
Chromosome

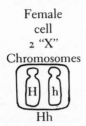

Hh

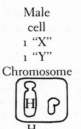

H__

Mother carries the recessive gene "h" in the hidden state on one of her "X" chromosomes. The dominant "H" is present on her other "X" chromosome and is responsible for her normal blood clotting characteristics.

Father carries only one gene ("H") for this trait since his "Y" chromosome cannot carry any genetic information about blood clotting factors associated with hemophilia.

Mother's ova will carry either "H" or "h"

Father's sperm will carry either an "X" or a "Y" chromosome so $\frac{1}{2}$ will carry an "H" gene and $\frac{1}{2}$ will carry no relevant gene.

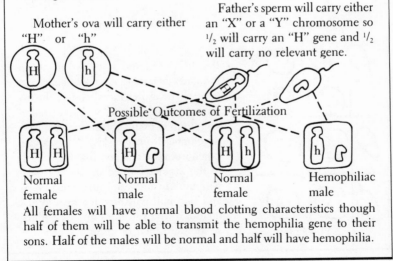

Possible Outcomes of Fertilization

Normal female | Normal male | Normal female | Hemophiliac male

All females will have normal blood clotting characteristics though half of them will be able to transmit the hemophilia gene to their sons. Half of the males will be normal and half will have hemophilia.

Illustration of the inheritance of cystic fibrosis, a disorder caused by a recessive gene located on an autosome (one of the 44 chromosomes in the human cell not classified as sex chromosomes), when neither parent is afflicted with the disease.

The man and the woman both must carry the same recessive gene for cystic fibrosis "c." The recessive genes are hidden by the dominant, "normal" gene "C" which prevents either of the parents from showing any symptoms of cystic fibrosis.

One gene from each gene pair is distributed to every sperm and every ovum.

Cc                    Cc

Sperm              Ovum
production      production

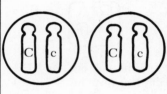

If a "c" sperm and a "c" ovum join to form a zygote, there is no dominant gene to mask the expression; therefore the information coded in "c" will produce the symptoms of cystic fibrosis in the person developing from that zygote. It can be noted from the diagram that one fourth of the zygotes formed in this marriage will result in children with cystic fibrosis and one half will produce healthy individuals who can transmit cystic fibrosis genes to their offspring.

ZYGOTES

*Illustration of the inheritance of Huntington's chorea, a disorder caused by a dominant gene.*

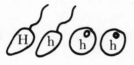

Hh
father

hh
mother

The father carries the dominant gene ("H") for Huntington's chorea but because of the nature of the disease he does not exhibit the symptoms until after he reaches sexual maturity. The mother need not carry any genes for the disorder. Since this disease is not sex-linked, the reciprocal situation (mother with "H") would be equally probable.

In this case one half of the sperm produced by the father would carry the "H" gene.

Since one half of all zygotes will receive an "H" from the father, the probability that a child will be born with this disorder is 50%.

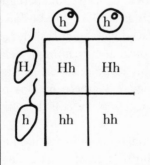

Persons with Huntington's chorea.

Persons without the disorder.

time with a spectrum of complications ranging from lack of bladder and bowel control to severe mental retardation.

## Genetic Counseling and Screening

Strategies are being developed to give prospective parents warning concerning their chances of having children with one of these genetic disorders and the possibilities for preventing that from happening (see chart).

In some cases examination of the family history can yield valuable information as to their chances. For example, if one or both of the prospective parents have relatives with cystic fibrosis, there is a much higher than average risk that one of their children will have the disorder. Genetic counselors are trained to work with the couple as they try to assimilate this new knowledge and make decisions about reproduction.

Until the development of prenatal diagnosis, couples who were at risk for genetic disease in their children had two basic options. They could either conceive and "take a chance" on having a child with a genetic anomaly, or they could decide not to conceive, but perhaps to adopt. Where feasible, techniques such as AID might also have lessened their risk. Many couples today faced with such knowledge still decide to adopt rather than to begin a pregnancy that might yield a child with a genetic anomaly. Since only some disorders can be diagnosed with any accuracy during pregnancy, these options of "taking a chance" or refraining from conception are still the only options in some cases.

But for some, the advent of prenatal diagnosis presents another option: selective abortion to prevent the birth of an affected child.

Prenatal screening is accomplished in several ways. In a few cases

## A. Interventions Yielding Information that may Prompt Decisions to Limit Conception or Control its Outcome

1. Genetic Counseling:
   —family history, blood tests, etc., which may prompt couple to avoid conception or to seek further testing upon conception.

2. Amniocentesis:
   —sampling fetal fluids to detect sex, chromosome make-up, or metabolic disorders.

3. Ultrasonic Scanning:
   —sound wave analysis of uterine contents to detect twins, observe body shape, size, etc.

4. X-rays:
   —analysis of uterine contents, as above.

5. Fetoscopy:
   —fiber optic probe of the uterus, with possible sampling of fetal blood.

## B. Interventions which Alter the Outcome of Conception

1. Abortion.

2. Sperm Sexing:
   —treatment of semen to increase probability of a particular sex of fetus.

3. Treatment of fetal abnormality in utero:
   —blood transfusions, drugs, surgery, etc., that can sometimes be performed directly on the fetus in utero.

blood tests of the parents may yield information. Carriers of recessive genes for sickle cell anemia and Tay-Sachs disease can be identified this way, and mothers carrying fetuses with neural tube disorders will have unique chemicals (or chemical levels) in their blood, as we saw in chapter 6. Additional tests for other disorders are being developed.

Most prenatal screening, however, involves amniocentesis. As indicated in the description of amniocentesis in chapter 6, amniocentesis involves withdrawing some fluid from the amniotic sac during pregnancy. Both the fluid itself and the cells that it contains from the fetus can then be analyzed. This gives the basis for an actual diagnosis of disease in the fetus as it is developing.

When the cells of the fetus are "karyotyped" (a map of the chromosome is drawn), it is possible to know both whether there are major chromosomal disorders and whether the child is a boy or a girl. Thus, the major chromosomal disorders and the sex-linked "monogenic" disorders can be diagnosed this way. It should be remembered, however, that in the case of the chromosomal disorder, the diagnosis only indicates the presence or absence of chromosomal material; it does not tell how severe the expression of the disorder will be. In the case of sex-linked disorders, the diagnosis tells the sex of the child. If it is a boy, there is still a fifty-fifty chance that it is not afflicted with the genetic disorder. Thus, parents who choose to abort following prenatal diagnosis may be taking the risk of aborting a child who is normal or who may live a relatively normal life.

Because the karyotype shows the sex of the child, some parents have sought prenatal screening to assist them in controlling the sex of their offspring. As a rule, physicians are reluctant to allow these procedures to be used for any reason not directly involving health risk. The use of prenatal diagnosis to select a boy or a girl has been considered a "trivial" reason, not sufficient to justify the test. However, under certain circumstances, knowing the sex of the child might not be a trivial reason. Certainly in the case of sex-linked disorders, the sex of the child is a clue to possible disease and would not be a "trivial" reason for seeking prenatal diagnosis.

For certain other "monogenic" disorders and for the "multifactorial" disorders, specific tests have been developed to locate fetuses with the disease. Often, these are biochemical tests that assess the presence or level of different chemicals in the amniotic fluid (as we saw with spina bifida and the "neural tube" disorders). Once again, as with many genetic disorders, a diagnosis that a neural tube defect is present does not provide a full prognosis of the child's possibilities. If ultrasonic screening indicates that the child has anencephaly, the prospects are quite bleak—minimal brain development and death occurring usually within weeks. But if the disorder is spina bifida, we cannot know for sure how severe the disabilities will be: some children with spina bifida are mentally retarded, but some are not; most are paralyzed from the waist down, but some will have other physical problems as well. Therefore, as is the case with Down's syndrome, with sex-linked disorders, and with monogenic disorders such as sickle-cell anemia, parents who decide to abort a child with a neural tube defect cannot be given absolute assurance that they are aborting a child with massive difficulties. Some might lead almost normal lives.

Because of these many capabilities for prenatal screening, amniocentesis is recommended for certain groups of people considered to be at high risk of passing on a genetic disease: those with family histories of genetic disorder, people from ethnic groups with a high incidence of an otherwise rare genetic disease (e.g., sickle cell anemia in the Black and Latin communities, cystic fibrosis in Caucasian groups, and Tay-Sachs disease among Ashkenazi Jews), and women over the age of thirty-five, who have an increased risk of major chromosomal abnormalities.

Some couples, even knowing that they are at risk, may choose to forego any prenatal diagnosis. Or they may choose to have the diagnosis but without any intention of having an abortion if the results are "positive" and the fetus is affected. Simply knowing mid-way through pregnancy whether the child is healthy or not would give parents time to prepare for the birth of a child with special needs. At present, however, the major use of prenatal diagnosis is linked to the possibility of selective abortion to prevent the birth of a child with a genetic anomaly.

It does not require great intuition to predict ever-expanding knowledge and capabilities in these areas. More genetic diseases will become accessible to diagnosis during the prenatal period. More blood tests will be developed to detect carriers of recessive genes for genetic diseases. New gene splicing techniques promise to revolutionize the detection of genetic disorders by utilizing DNA "tracers" to locate normal genes linked to defective genes. Some day, we may be able to replace those genes with techniques gleaned from "recombinant DNA" research. There is hope that someday we will be able to repair genes as well as to screen couples at risk and offer the possibilities of abortion or avoiding conception.

Therapies for many genetic disorders will also become more available. Today, for example, researchers are close to a successful treatment for sickle-cell anemia. Since this recessive disorder afflicts 250 out of every 100,000 blacks in this country, such a treatment would be a major breakthrough in relieving suffering for afflicted persons and their families. Treatments during pregnancy are beginning for some disorders, including dietary restrictions on the mother or even surgery in utero. These developments will add new dimensions to the process of making decisions about prenatal diagnosis and selective abortion—for example, should we abort if there is a treatment available instead?

Finally, our ability to keep human beings alive during both the early and late stages of fetal development raises ethical issues at the same time that it provides the possibility of new interventions. For example, in vitro fertilization is the key to gene splicing and "repair" of genetic material through recombinant DNA techniques. As in vitro fertilization becomes more dependable in humans, it promises not only to open up the range of people involved in child-bearing (as discussed in Part I), but also to offer new possibilities for controlling genetic disease. Similarly, neonatal intensive care has now progressed to the point that very premature infants can be saved. Some day we may be able routinely to save premature infants as young as eighteen or nineteen weeks of gestation. This would be close to the time that selective abortion is done following prenatal diagnosis through amniocentesis. How will we decide about abortion or subsequent management of the newborn? These developments raise difficult ethical questions and push us toward deep theological reflection about the nature of our power over genetic disease.

Now that we have looked at a specific case involving questions of "quality control" and have examined a bit of the scientific data about genetic disease and possible interventions, we turn to an exercise designed to help us think about different kinds of disease and what interventions might be appropriate. Following that, we will raise some theological questions about the meaning of illness, suffering, disappointment, and healing in human life.

# CHAPTER 8

# On Having
# "Normal" Children

What expectations do parents have about "normal" children? Are these hopes legitimate? How do we connect our efforts at eliminating genetic anomalies with our response to people in our communities who have those anomalies?

The spina bifida case study in chapter 6 opened up the issues in a particular case and enabled us to explore those particular circumstances. Now we consider a greater variety of possible problems. New technologies for screening and counseling parents about genetics provide many opportunities for parents to choose whether a life with certain characteristics is to be conceived or born. This is an awesome responsibility.

As you read this exercise, try to identify your values and attitudes and locate the factors that seem most critical to your response. One of these cases has to do with prenatal diagnosis and selective abortion. Abortion remains very controversial in our society. While you will be challenged here to think about whether you can accept selective abortion, we propose that you focus primarily on the question of normality and expectations about the qualities of children, not on the abortion question per se.

The suggested procedure for completing this exercise is this:

1. Read and consider these vignettes and the questions following them individually. Save the "discussion questions" at the end for small group discussion.
2. Share your initial reflections and concerns in small groups.
3. Next, turn to the discussion questions and examine them in small groups.
4. Finally, the whole group should process the issues and try to identify central questions that you want to explore in greater depth. In particular, we urge you to focus on *which factors* make a difference in your judgment. If you think a child should be born in one circumstance but not in another, what makes the difference?

### PHILLIP AND ALICE

Phillip and Alice have been married twenty years and have three children, two in college. Now forty-one years old, Alice unexpectedly becomes pregnant. On the advice of her doctor, she undergoes amniocentesis since there is a significant risk (c. 2–3%) that the fetus might have Down's syndrome. The prenatal test shows that the fetus does *not* have Down's syndrome. But it *does* have a disorder called Hurler's syndrome.

In Hurler's syndrome, the baby appears normal for a few months. Then its skin coarsens and its joints stiffen. The child does not grow normally. Hydrocephalus ("water on the brain"—the accumulation of fluid in the head) can develop, and the child will typically be mentally retarded. Children with Hurler's syndrome usually succumb to respiratory or heart disease and die by age 10.

Do you think this child should be born? Why or why not?

Now ask whether changed circumstances would alter your decision. Suppose, for example, that:

- Researchers report new evidence and treatment that can reduce or even eliminate the effects of mental retardation. The child will still die early.
- Better research and statistical studies show that the life expectancy is actually closer to fifteen or twenty years . . . or is closer to four or five years.
- Phillip has just been laid off from his autoworker's job, Alice makes barely enough to support ongoing family expenses, and family savings and life insurance are presently committed to college education for the three children they already have. State resources for the handicapped have been drastically reduced in recent budget cuts.
- One of the other children already has a costly physical handicap requiring extra assistance in equipment and physical therapy, as well as a great deal of care and understanding by the family.
- Phillip and Alice lost a fourth child to "sudden infant death syndrome" fourteen years ago. (In SIDS, sometimes called "crib death," an apparently normal and healthy child dies in its sleep.) Alice was depressed for two years following that episode.
- Phillip and Alice have experienced strains in their marriage in recent years. This unexpected pregnancy is a major source of stress in their increasingly shaky relationship.
- The pregnancy was planned and they are eager to have another child, having missed the "patter of little feet" since their three became teenagers.

As you think about these possible variations in their circumstances, try to isolate which factors seem to you to justify a decision for abortion, and which do not. Can you formulate any general rules or guidelines about what things seem most important in considering the quality of life of the child and the parents' circumstances?

## MICHAEL AND PATRICIA

Michael and Patricia gave birth to a child with cystic fibrosis three years ago. While Tommy can do most things that other three year olds do, his continued health is dependent on a rigid family routine: several sessions a day of thumping gently on his back to loosen the mucus in his lungs, regular checks with the doctor, and occasional hospitalization or medication. Although the family employs a maid to assist with household duties and child care, attending to Tommy's needs is an important feature of the family's life together. Patricia is constantly on guard for Tommy's health, and feels confined to the home and the needs of their child.

Michael wants another child. He feels a bit "cheated" by God that his son will never be physically very strong or active. He refuses even to think about Tommy's eventual death. He would like to have another son—or perhaps a little girl.

Patricia, however, is not so sure. Having learned that cystic fibrosis is a recessive inherited disease, she is afraid that the second child might also have the disease. She is angry with God for Tommy's illness. She prays hard for some "miracle" of healing, but deep in her heart she doesn't believe that it will happen, and she sees herself saddled with a sick child for twenty years, only to watch him die a terrible death at the end. Even with household help, the idea of having two such children seems more than she can bear.

Michael and Patricia learn that their chances of having another child with cystic fibrosis are one out of four. A prenatal test is being developed, but it is not yet considered "routine," so there is a chance of a "false negative"—of being told that the child is healthy when in fact it will have the disease. Michael and Patricia have a number of options:

- They can adopt and avoid all risks.
- They can conceive and have the child.
- They can conceive and try the prenatal testing.
- They can try AID, so as to avoid the risk.

Which of these options would you propose for them? Why? Which factors in the case as described above seem most important?

Now once again, ask whether changes in the circumstances would change your judgment:

- The child will not only have respiratory problems, but will be severely retarded with little likelihood of mental capacities progressing beyond an infant level.
- The child will not die before adulthood, but will have stunted physical growth and will be infertile. They will have to deal with questions about sexuality, marriage, and parenthood for this child.
- Michael and Patricia have two older children whose social adjustment and psychological development have been hampered by the attention given to Tommy. The daughter in particular "hates" Tommy and is a very jealous and possessive child.
- The couple has just given birth to an infant with Down's syndrome ("mongolism").
- Financial pressures will soon necessitate abandoning the help of the maid; Patricia already has very few friends because of the demands on her time at home.

## DISCUSSION QUESTIONS

1. What kind of expectations should parents legitimately have about the health and normality of prospective children? Should they resist the birth of "defective" children?
2. What is normal human life? What do our responses to these vignettes and our general reaction to "abnormal" people suggest about our concepts of normality and our acceptance of disabled persons?
3. When parents give birth to "defective" offspring, what response and assistance ought to be given by the family? by friends and the church? by social and public agencies?
4. What role should family circumstances and relationships play in assessing responses to handicapped children?
5. Should we interfere in natural processes? Is it fair when a family must suddenly confront the possibility of a severely handicapped child? What does our understanding of biblical faith suggest about appropriate responses?

# CHAPTER 9

---

# On Illness, Disappointment, Suffering and Healing

---

From the beginning, Christian faith has celebrated God's power in healing. The four gospels record many stories of the healing acts of Jesus. Sometimes faith leads to healing, and sometimes healing evokes faith. But sometimes there are grave problems in the relation between faith and healing.

Thus John the Baptist sent messengers from prison to ask of Jesus: "Are you he who is to come, or shall we look for another?" Jesus replied: "Go and tell John what you hear and see: the blind receive their sight and the lame walk, lepers are cleansed and the deaf hear, and the dead are raised up, and the poor have good news preached to them. And blessed is he who takes no offense at me" (Matt. 11:2–6). It was a stirring answer. But it did not release John from prison or save him from Herod's execution. God's healing power does not always immediately overcome injustice, pain, and death.

People long for healing only because there is sickness. Human

73

life is frail, prone to illness, directed toward health. Even a good creation includes pain, disappointment, frustration. If some pain serves a purpose and if some suffering ennobles life, other pain and suffering make purposeful living exceedingly difficult.

Genetically related illnesses are likely to raise especially troubling questions. They are sometimes extremely severe, as noted above. At their worst, they can distort much of what we find meaningful in human selfhood. They are, or seem to be, absurd, in the literal sense of that word—capricious, irrational intrusions into the rational and purposeful nature of life. They lead suffering people and the suffering parents and friends of afflicted people to ask: "Why did this happen to me?" "Am I being punished?" "Am I guilty?" "How can a loving God permit this pain?" "What can I do?" "How can I cope with this disappointment?" Or, as we saw in the case of Michael and Patricia in the last chapter, people respond with anger toward God—possibly even with loss of faith.

These age-old questions about pain and evil arise in every generation. They are not likely to find new answers in our generation. And we are not likely to be satisfied with answers from the past. Yet people have found in the heritage of faith resources for living with pain and frustration. We can here look at four biblical themes that give us resources for understanding the role of illness, suffering, disappointment, and healing in Christian life.

## HUMAN FRAILTY: ISAIAH 40:6–8; MATTHEW 7:28–30

A starting point is the insight of the prophet Isaiah of the Exile:

> All flesh is grass,
>     and all its beauty is like the flower of the field.
> The grass withers, the flower fades,
>     when the breath of the Lord blows upon it;
>     surely the people is grass.

> The grass withers, the flower fades;
>     but the word of our God will stand forever.
>
> (Isa. 40:6–8)

It is not as though faith starts with a confidence that life will turn out well, then unexpectedly runs into the discovery of disappointment. On the contrary, faith starts with the realization that we human beings are not self-sufficient, that all life includes pain, that all life moves toward death. One of the messages of faith is that we human beings, even at our strongest, cannot create or sustain ourselves. Our mortal lives are completed only in God.

Jesus rang an interesting change on the words of Isaiah that he knew so well:

> Consider the lilies of the field, how they grow; they neither toil nor spin; yet I tell you, even Solomon in all his glory was not arrayed like one of these. But if God so clothes the grass of the field, which today is alive and tomorrow is thrown into the oven, will God not much more clothe you, O ye of little faith?
>
> (Matt. 6:28–30)

For Jesus, as for Isaiah, the grass becomes a sign of the frailty of all created things, including ourselves, even as it is a sign reminding us of the eternal power of God. Our God, who gives such glory to the brief lives of grass and flowers, cares for us also in our weakness as well as our strength. Indeed, it is a frequent theme in the Bible that we are to be entirely dependent on God for our sustenance and strength.

## GOD AND UNDESERVED SUFFERING: EXODUS 20:5–6; THE BOOK OF JOB

Some illness and pain is the result of human ignorance, folly, and sin. Often the Bible—and our own common sense—tells us that rash and evil deeds bring destructive consequences. In the Ten Commandments we read: "I the Lord your God am a jealous God, visiting the iniquity of the fathers upon the children to the third and fourth generation of those who hate me, but showing steadfast love to thousands of those who love me and keep my commandments" (Exod. 20:5–6). We may complain that some must suffer for wrongs done by others. But life is so bound up with life that all of us share in the rewards and punishments brought upon the common life by others. Nowhere is this better illustrated than in the effects of environmental pollution on our health.

Out of that true insight has come the false belief that suffering, in general, is a punishment for sin. Something in most of us (and something in the ancient Hebrew people) seems to *want* to believe that. It would help us to make sense out of our lives if it were true (cf., Ron Green, *Religious Reason*). Yet biblical testimony emphatically denies it.

When Job in his agony wanted human sympathy, his friends moralistically told him that his sufferings were his own fault. They tediously argued that he was somehow to blame and that his very denials were part of his guilt. For their efforts they have earned the ironic title of "Job's comforters." But God, according to the record, preferred Job's angry complaints against God to the friends' unctuous insistence that Job was irreverent: "My wrath is kindled against you . . . for you have not spoken of me what is right, as my servant Job has" (Job 42:7).

From this, we learn an important lesson about dealing with illness and suffering: when we suffer, in our own pain or in the sufferings of others, one sign of faith may be anger against God. Anger is not the only expression of faith. But there are days when it may be the most genuine thing we do. Sometimes it is good to complain!

SUFFERING AND GRACE: JOHN 9:1–12;
2 CORINTHIANS 12:7–10

Jesus refuted directly the belief that suffering is always the result of sin. On one occasion his disciples, seeing a blind man, asked the familiar question, "Rabbi, who sinned, this man or his parents, that he was born blind?" Jesus answered them: "It was not that this man sinned, or his parents, but that the works of God might be made manifest in him" (John 9:2–3). With that, Jesus healed the man, restoring his sight.

The text does not require us to believe that God deliberately made this man blind from birth in order that some day Jesus could come along and give him sight. The point is that illness and suffering are opportunities for God's mercy, working through persons who heal and support those in pain. Although it is not *wrong* to ask, "Why does God permit suffering?" and although we sometimes cannot hold back from asking that question, we may never get an answer. But there is another question more purposeful: How does the reality of suffering give opportunity for the works of God to become manifest in human life—through us and through all those who help and sustain life in weakness and in pain?

The story of this blind man had what we tend to call a "happy ending": the man received his sight. (There was some grumbling, of course, because Jesus healed on the Sabbath, but the joy surely outweighed the discontent.) However, not all sickness ends in health and not all pain finds relief. Nor is *our* definition of a "happy ending" necessarily the same as God's, as Paul's struggle with his own ailment shows.

The Apostle Paul had a physical ailment, which he called a "thorn in the flesh." As he tells the story, "Three times I besought the Lord about this, that it should leave me; but he said to me, 'My grace is sufficient for you, for my power is made perfect in weakness' " (2 Cor. 12:8–9). It is hard enough to take the preliminary step of this faith-full understanding: to realize that God does not despise weakness. It is far harder, but very important, to take the next step: to appreciate that God's power has a special relation to our weakness, that God is with us in our weakness. This challenges us to reverse our usual understanding of "happy endings," to see that the best result to come out of illness or disappointment is not necessarily healing or removal of the problem. As Paul himself puts it, "That is why, for Christ's sake, I delight in weaknesses, in insults, in hardships, in persecutions, in difficulties. For when I am weak, then I am strong" (2 Cor. 12:10).

## SUFFERING AND INCARNATION: ISAIAH 52:13–53:12; 1 CORINTHIANS 1:18–31

The most audacious step of all is the recognition that when God most decisively enters human life, God does so in the person of a Suffering Servant and a crucified Savior.

The same Isaiah of the Exile who declared that "all flesh is grass" tells of the mysterious Suffering Servant of God, one whose appearance was "marred, beyond human semblance," one who "had no form or comeliness that we should look at him" (Isa. 52:14; 53:2).

> He was despised and rejected by men;
> a man of sorrows, and acquainted with grief.
>                                    (Isa. 53:3)

As this book is being written, the movie and play versions of "The Elephant Man" are very popular. Here is the story of one who was "marred, beyond human semblance" and yet brought the gift of grace to others. Thus we see that any person, no matter how "despised and rejected" in the human community, can be a vehicle for God's redemptive activity.

Of course, the Christian church has from the beginning identified the Suffering Servant with Jesus Christ, who becomes the norm and paradigm for other sufferers who bring grace. And Paul exults in the wonder of it all: "For the word of the cross is folly to those who are perishing, but to us who are being saved, it is the power of God" (1 Cor. 1:18). And he continues, "God chose what is foolish in the world to shame the wise, God chose what is weak in the world, even things that are not, to bring to nothing things that are, so that no human being might boast in the presence of God" (1 Cor. 1:27–29).

We can still rejoice in human beauty and intelligence and skill, all signs of God's creativity. But for God's crucial, redemptive deed we look at the Suffering Servant, at power made perfect in weakness.

No one of us knows entirely how to realize this faith in life. It does not mean that we should want to suffer or deliberately seek illness. Jesus in Gethsemane prayed that he might be spared the cup of suffering. But when suffering comes to us and those we care about, we can sometimes recognize God's grace and power. And we know that suffering is not *always* something to be gotten rid of as quickly as possible. The Christian gospel does not answer the question, "Why does God permit suffering?" Christian faith neither prescribes nor guarantees any answer to that question, though many have been proffered. Christian faith does affirm that God works through suffering as well as through healing. It does say that God joins us in our suffering, shares our burdens, and sustains us in our weakness.

## QUESTIONS FOR DISCUSSION

1. Why was God's "wrath . . . kindled" against the three friends of Job? Do you agree with the statement that "one sign of faith may be anger against God"?
2. Why was it necessary for Jesus to insist that the blindness of the man (John 9) was due neither to his nor to his parents' sin? Is it still necessary for the church to assure people on this point?

3. Novelists sometimes create characters known as "Christ-figures"—that is, characters who, though very different from Christ, nevertheless remind readers of Christ in certain symbolically important ways. In William Faulkner's *The Sound and the Fury*, Benjy the idiot is a Christ-figure. Do you think such a literary device can be an act of reverence?

4. Are there ways in which churches can appreciate disabled persons, not simply as objects of compassion, but as ministers to the presumably "normal?"

5. Have you ever seen, in contemporary life, signs of God's power "made perfect in weakness"?

6. Evaluate each of the following actions, which might take place in some real or imaginable future:

   a. A skilled basketball player never quite achieved greatness, and thinks he was not tall enough. Wanting his sons to be great players, he (1) seeks a tall woman to be his wife, or (2) consults a genetic counselor on ways to get tall sons.

   b. A government, worried about national security, starts a research program with the hope of producing a future generation with exceptional mental and physical abilities for warfare.

   c. A foundation considers funding a genetic research program with the aim of manipulating DNA so as to eliminate some serious genetic diseases.

   d. A state declares that no medical or support services will be funded by the state for families who deliberately choose to bring into the world children with genetic handicaps.

   e. A church promises healing to all who enter.

# PART III

## *Making Decisions and Getting Involved*

# CHAPTER 10

---

# Making Ethical
# Decisions

---

We have now looked at some cases and exercises to stimulate our thinking, provided some scientific background to clarify possibilities, and offered some theological reflections to help us sort out some of the "deep structure" of faith and what it would say to the issues at hand. Throughout, we have stressed that there is no simple answer, and that the Bible is not a rule-book for making decisions about responsible Christian living.

But how, then, *do* we make decisions? What do we do with the scientific data and the theological reflections? How do we use the "gut-level" responses we may have had to the case studies, especially as we looked at them from the perspective of one or another of the roles or players involved? The purpose of this chapter is to suggest a method for putting these factors together. We want to stress that this is "a" method; it is not the only one nor necessarily the best one. But it may help Christians who want to struggle responsibly with these decisions today.

## On Rules and Situations

First, it may be helpful to say a word about rules and situations. Many Christians have held firmly to the view that there are some rules for Christian living—that adultery is wrong, that innocent people should not be killed, and so on. If "loving our neighbor as ourselves" is central to Christian ethics, then it seems as though love—if it is genuine love—will issue in some rules about things that ought never to be done. Rape, torture, and perhaps a small handful of other acts seem so evil that they could never be justified in the name of "love."

But this stress on the rules for behavior that issue out of love has been challenged by those who take a "situational" or "contextual" approach to ethics. "Everything depends on the situation," some claim. Is adultery always wrong? Suppose a woman is interned in a Nazi concentration camp, and the only way she can be freed is to get pregnant. Is it wrong for her to commit adultery in order to escape and return to her family? There always seem to be such "borderline" cases that throw our hard-held rules into question. Perhaps rape and torture are always wrong, but very few other things seem to have this "absolute" quality when we really think about it.

In the wake of the "situation ethics debate," many Christians became confused. Are there rules or not? If so, what are they? If not, then of what use is the Bible? Is everything just left up to intuition? What happens when good people disagree in their intuitions?

We cannot answer all of these questions in full here. But we can suggest some helpful hints.

First, the "rules vs. situation" debate may be a "misplaced" debate. Rules imply situations, and situations imply rules. For example, "adultery is always wrong" is a rule that is written about a specific situation—namely, the situation of adultery. The question may simply be how broadly to write the rule: shall we say, "adultery is always wrong" or shall we say "adultery is always wrong, except where it supports the basic value of bringing covenanted partners back together"? In short, "exceptions" to rules may just mean that we have not written our rules with enough care.

If we are careful about rules, we discover that underneath the specific rules about which we argue, there are more general rules (sometimes called basic principles) about which we all agree. In the example above, for instance, the more general principle that underlies the rule against adultery is the principle to keep covenant. Most of the time this general principle would prohibit adultery, but sometimes it seems to permit it. Disagreements about specific rules for behavior do not imply that we disagree about the general principles (such as respect for persons, or justice) that underlie those specific rules.

Thus, the first thing that we can say about rules and situations is that much depends on whether we are talking about specific rules or general principles. Specific rules tend to have exceptions, or "boundary" situations where it is difficult to apply them. But general principles may not have this same uncertain character. Everyone would agree that "justice" should be done. What we might not agree upon is how to define justice or to work it out in a specific case, or how to draw specific rules from the general principle. For instance, does "justice" require giving in accord with need, giving in accord with accomplishment, or some combination of these two specific rules?

Second, just as rules are written for situations, and much depends on how narrowly the rule is written, so situations point to rules. Indeed, situations cannot be described and understood unless we have some rules. This point is perhaps harder to grasp, but it is just as crucial for our purposes.

Perhaps adultery is not always wrong. But what is adultery? How do we know whether an act constitutes an act of "adultery" or not? We do not even know when to describe the act of making love as "adultery" unless we have some rules about love-making, marriage, fidelity, and so on. Thus, to say that "everything depends on the situation" is not helpful, because we don't know what the "situation" is unless we have some rules (or norms) that we can use to help us define and describe our world.

Moreover, in the example above, the focus on adultery seems misplaced. The *first* thing to say about this situation is not that it is a situation of adultery, but that it is a situation of horrifying and unacceptable oppression. How we describe a situation has much to do with what our options are for response. The description itself depends on our perspective or "vision"—on our loyalties, our past experiences, our faith. Thus, in addition to rules and situations, we have basic faith affirmations—for instance, that God intends us to live in community, or that God values all persons equally. These affect the very way we see the situation.

When we come to define a situation, then, we already have in mind some basic principles. Our faith affirmation that God intends us to live as equal persons in community is what tells us that the Nazi experience was an experience of oppression. The act of adultery must be assessed within this definition of the situation. Situations do not simply appear; we *view* them. And as we view them, we bring to bear the perspectives of our faith and of our life experience.

We can illustrate both of these points with examples drawn from new genetic technologies. We generally respect the self-determination of human beings by having a rule to obtain "informed consent" before operating or experimenting on someone. But how can we apply this rule to the case of the unborn—or even the "unconceived"? Does this new "situation" show that no rules are absolute, since the rule for informed consent does not apply? We would respond that while the rule does not apply, the basic principle that gave rise to the rule still does apply. Hence, the question is: How can we respect the persons that we are thinking of bringing into being? Since these persons cannot give or refuse consent on their own behalf, are there some other ways that we can protect their integrity and their personhood? When thinking about the risks to a child of being conceived by in vitro fertilization, for example, is it enough to permit the parents to make decisions about the acceptable level of risk? Provided they are operating within a covenant concept of family, and are willing to raise and nurture the child even if it is not "perfect," it is probably sufficient to permit them this decision. This would explain why many Christians responding to the exercise in chapter 4 become uncomfortable the further away from a covenanted family we move in our scenarios about in vitro fertilization. The rule of getting informed consent needs to be understood within the context of the deeper principles (of respect for persons, and of covenant) that give meaning to the rule. It is these principles that help us to decide what rules to have and when to break them.

Similarly, our discussion about "quality control" shows how important basic principles and faith affirmations are for helping us determine what the "situation" is and how to respond to it. Many times when a child is born with a genetic anomaly, friends and family (and professional helpers) respond by saying, "Oh, what a tragedy." We noted that parents sometimes respond by asking, "Why did God do this to me? What have I done?" In these responses is a definition of the situation that depends on a normative view of humanity. The situation is defined as "tragic" or "disappointing" because of our assumptions and expectations about how human life should be. Once we have defined the situation as "tragic," a whole new set of rules for response gets brought into play—we respond with the rules we know for coping with tragedy. We cry, we get angry with God, we "pick up the pieces and start again," we work to rid ourselves of the problem. All of these are the responses that we "expect" and accept in the face of tragedy. Thus, the definition of the situation as tragedy brings to bear a lot of rules for how to respond. Different cultures or sub-cultures might define the situation differently, and therefore might respond to it differently.

Rules and situations are deeply intertwined, therefore. There are no rules without implicit ideas about situations, and there are no "situations" that can be defined and responded to apart from rules. The question is how we write our rules, and how they reflect those more basic ethical principles upon which they rest. Moreover, equally important with rules and situations themselves are the loyalties and faith affirmations that we bring to any situation and that help us to define what the "situation" is—to highlight some aspects of it as more important than others, and to describe it in a way that helps us to know what kinds of rules might apply.

## Ethical Principles and the "Burden of Proof"

This recognition has led some contemporary Christian ethicists to seek a way of making decisions that is not bound by rules but also does not ignore the basic ethical principles that give rise to rules. One way of doing this is the following:

Our Christian faith gives us certain basic affirmations about God and about the meaning and purpose of human life. Above, we called this the "deep structure" of faith or of Scripture. For example, as we noted in chapters 3, 5, and 9, biblical faith tells us that families are important but that they are not limited by genetics; that power is problematic and must be used within the boundaries of our existence as "creatures" exercising responsible stewardship rather than striving to "be God"; and that suffering can be a vehicle for God's grace and presence in the world. Taken alone, these basic affirmations arising from our biblical faith do not tell us what to do. But they do give a certain direction to our lives.

From them, for instance, we might derive the following kinds of basic principles: (1) Things that support and nurture community are generally within God's intentions for human life, and things that undermine community violate God's intentions. (2) Power must be used in the service of God and humanity, not for our own self-aggrandizement. (3) Suffering is not necessarily to be considered an "evil" and to be overcome at any cost. Each of these principles might then give rise to all sorts of specific rules—e.g., rules about fidelity in marriage, responsibility for child-rearing, balancing access to the goods of the world, and so on. There will always be much disagreement about the specific nature of the rules; however, if we can agree on the basic principles, we have a place where fruitful argument can take place.

Finally, then, to these basic affirmations and the kinds of rules that we derive from them, we can add the notion of "burden of proof." The "burden of proof" is on anyone who proposes an action that does not seem to be in line with these basic principles and faith affirmations. For example, since life is a good, anyone who proposes to take life must justify that action; since fidelity is a requirement of human life, anyone who proposes to violate fidelity must justify that action; and so on. Actions that support the basic affirmations are presumptively correct; actions that violate those affirmations are presumptively wrong and must be justified.

## Justifying Actions

But now comes the crucial question: what does it take to justify such an action? Can I simply say that my "situation" is different from yours, and that therefore the action is justified in my case, even though it would violate a rule or principle? This is what the situational ethics approach seems to do. We are told that some situations justify breaking rules, but we are left unsure as to just which aspects of the situation count and are sufficient to justify breaking the rules.

At this point, we ask you to remember some of the cases and exercises, and the questions you discussed following them. Each case is a different "situation." As you thought about them, and discussed them, did any patterns appear in the kinds of things that you thought "made a difference" in your assessment of the case? Think of chapter 8 in particular. As you thought about the situations of Phillip and Alice or Michael and Patricia, which *changes* in the situation would have made you answer differently? Are there some circumstances under which abortion or taking or refusing risks seem more justifiable than they would in other circumstances? If so, what are these factors that make a difference, and why?

Some factors are "morally relevant." They do make a difference in the moral assessment of the situation. They change the justification of the action contemplated. We all have an intuitive grasp of this concept, and we know that we use it in our everyday decision-making, but *why?* And *which* factors count like this? These are the crucial questions.

There is no simple answer to what factors are morally relevant and what factors are not. Good people, and Christians, will disagree. In general, though, we can say at least this much about which factors should make a difference and which ones should not: *The things that make a difference are those that affect how a basic affirmation or principle might apply.* In short, "morally relevant" factors are things that are justified by basic principles.

For example, many of you reading this volume probably intuitively felt that the financial situation of Phillip and Alice "made a difference." Others of you probably did not think that money should count, especially where human life is involved. In order to know whether their financial status is morally relevant, and might make a difference, we have to ask how it affects basic principles and their application. Suppose we hold *covenant* to be a basic principle, as most Christians do. Then we want to say that actions that uphold covenant are generally right, and those that undermine covenant are generally wrong. But now the question would be, "How does their financial status affect their ability to covenant with each other, with God, and with their children?" Some might think that finances have nothing to do with covenant. But if it can be shown that lack of resources makes it extremely difficult, if not impossible, for them to be the kind of loving, covenanting parents that they wish to be, then their financial circumstances would seem to be relevant to the decision, because it would affect the enacting of a basic principle.

It is for this reason that we need clarity both about the empirical data and also about the theological affirmations that we as Christians might hold as basic convictions. The scientific and empirical data help us to understand what the impact of slight changes would be: will the child live five years, or twenty-five years? This might be "relevant" to the decision, and hence it is incumbent upon us to get the most accurate data we can.

At the same time, we do not know what use to make of the data, or whether it is indeed relevant, until we put it with our basic theological affirmations. Not until we agree that life is a good, that covenant extends beyond genetic families, and so on will we know what to do with the information that someone will have more or less life, or that this action will bring into a family someone who is not genetically related. We assess the meaning and the use of our scientific data by putting it within the framework of that "deep structure" gained through Christian faith.

## Character and Virtue

Part of our faith is the conviction that we are "made new" in Jesus Christ. Thus, in assessing what to do, it is not enough to ask for rules, or even to weigh scientific data in the light of faith affirmations. Since a change in who we are is central to our faith, we must also ask in what ways the options we have would affect our character and the kinds of people that we are and want to become.

In short, another problem with the "rules vs. situations" approach to ethics is that it seems to take every action as though it were a separate action that bears no relation to everything else in our lives. But we are ongoing. What we do affects who we are. It may be perfectly justifiable for me to tell a lie today. But if I lie over and over again, no matter how well each lie appears to be justified in the situation, at some point I must raise a question as to whether I have become a liar. One of the ways that I might decide what to do is by looking at how this action fits into the overall pattern of my life.

This is another way of saying that Christianity is at root a *story*, and that our stories count. Who am I, and who am I becoming? If I choose not to bear this child, what kind of person does that help me to become? If I seek pregnancy at any cost, what am I saying about myself and my image of myself? These would be alternative ways to put the kinds of questions we have been putting above. Instead of asking whether actions are right or wrong, we might ask about the development of Christian character.

This in turn leads us to ask about virtues in human life. Which qualities ought we to cultivate? Which motivations might we consider virtuous and which vicious? Can we even use these words anymore?

Again, Christians will disagree about specific virtues. But we share a basic story or faith saga (the Bible) that consists of many stories that help us to understand virtue. The story of Ruth does not give us direct guidelines for parenthood; but it does tell us something about the virtue of fidelity to those who are not related to us genetically. The story of Job does not tell us whether to institute massive screening programs designed to locate women at risk for neural tube defects. But it does tell us something about the kind of person God loves. An important part of our reflection about these ethical dilemmas would be the telling—and re-telling!—of our common story, and the effort to pull from that story some virtues that we might seek to build in our own lives.

An important set of virtues would be those that have to do with being "good parents." We have lots of images of what it means to be a good parent—a good provider, a careful listener, a resource for the child's growth, etc. Some of these images are changing—for example, the image of the "good mother" as one who stays home all day to be available to the children changes both as women seek fulfillment outside the home and as economic pressures force more and more families to have two wage-earners. As we live our life stories, we usually have in mind some image that we try to live up to, or some person that we emulate. Our decisions about using new genetic technologies will have to do in part with what our image of "good parenthood" is.

Of late, there has been a lot of talk about "responsible parenthood." An image is emerging about what makes parents "responsible." Even people who disagree about almost everything are beginning to agree about one thing: that responsible parenthood requires parents to avoid bringing into the world children with serious and preventable handicaps. People disagree about which actions are permissible to secure this end—e.g., whether to permit abortions and endorse new interventions such as in vitro fertilization as means to achieve the goal—but they do agree about the basic image of responsible parenthood.

Without necessarily disagreeing, we can raise some questions from a Christian perspective as to what "responsible parenthood" really means. Are we really operating out of a *Christian* perspective on responsibility and parenthood, or have we simply adopted the norms for responsibility of the secular society around us? Does responsibility in Christian perspective require that we avoid bringing into the world children with handicaps, or does it require that we nurture and care for those who enter our world with disabilities? This question merits attention by church groups everywhere.

## SUMMARY

The method we have been offering here could be summarized as follows:

1. The first task in confronting any situation is defining what that situation is. This is a two-fold process.
   a. It involves looking at the empirical data, and learning the basic "facts." What are the options available, the scientific data, the implications of each option?
   b. Of course, these are not simply "facts." As soon as we talk about implications, we are talking about the meaning of actions. We cannot talk about the meaning of actions without addressing theological questions. Is a couple contemplating a screening program that might end in abortion? There are "facts" here, to be sure— what the screening tests involve, what the rate of "false positives" might be, and so on. But more important are the definitions of the options available, and how these are understood. Thus, the second step in the process is to try to see how our basic faith affirmations and principles are helping us to define what the situation is.
2. This then pushes us to locate basic faith affirmations.
   a. First, we look for the basic principles that our faith gives us for assessing God's actions in the world and our response. Is life a gift? Is suffering to be avoided? We need some clarity about these issues.
   b. Then we ask how the "facts" of the case might affect our interpretation and application of these principles. How much life will a prospective child have? How much suffering? Which facts are "morally relevant" and would change our assessment? How do the facts affect our ability to live by the principles?

3. From this, we can assess whether an act is "presumptively" justified. Does it begin with the "burden of proof" against it, or with it? Does it cohere with our basic principles, or does it violate one or another of them?

4. If it violates them, and is not presumptively justified, can it be justified nonetheless? Can we show that although it seems to violate the principles, in fact those same principles support that action, given the "morally relevant" facts of the case? For example, could a basic principle of covenant support an action such as abortion of a defective child? While this action might not seem consonant with the principle under some circumstances, are there circumstances in the case that make the application of the principle stretch to include this action?

5. And finally, how will this action affect our story as Christian people? Is it consonant with Christian virtues? Does it "fit" into a coherent life style based on our "deep structure" of faith (those basic principles)?

This method does not give automatic answers to the difficult issues facing parents and concerned church people today. Christians will disagree about the use of new technologies and their implications for parenthood. How nice it would be if we had a simple rule-book that said, "do this" or "don't do that"! But human life is a complex mixture of freedom, responsibility, suffering, grace, and paradox. In the midst of that complexity, the method we offer is meant to provide a beginning for sorting out issues.

# CHAPTER 11

---

# Getting Involved—
# The Responsible Church

---

In chapter 10 we offered a beginning methodology for making ethical decisions in these difficult arenas. But now we ask another crucial question: How do we nurture and support the people who must make and live with these difficult decisions? What is the role of the church? What can we do to help?

One study of new genetic technologies and the church suggested three types of involvement for the local church:

1. SUPPORT: Churches can provide resources for families in crisis situations, a community life of faith, love, and acceptance, and pastoral counseling, prayer, and assistance in finding meaning in life.
2. EDUCATION: Study groups such as those using this workbook can assist members of the congregation and others to understand the complex issues involved and to clarify options for approaching ethical decisions.
3. ADVOCACY: Church groups can be involved as advocates on public policy issues related to new genetic technologies—for example, supporting legislation to aid disabled persons, examining proposals for

genetic screening programs in the neighborhood, and giving a "prophetic voice" to issues raised for parents and others.

These are only suggested avenues for involvement. The activities
chosen and initiatives taken by any congregation will reflect its
experiences, needs, and interests. As suggested at the outset, some
may be more interested in "exotic" technologies such as in vitro
fertilization and their impact on our understanding of the role
of parents. Others may wish to focus on decisions of parents
facing genetic counseling or screening. Support, education, and
advocacy are three possible modes of response to any issue or focus
chosen.

In this brief concluding chapter, we offer two discussion processes
to help the study group identify the particular needs, interests, and
resources of its own local congregation so that it may begin to answer
the question, "What can we do?" Either or both of these discussion
processes could be used. They will suggest possibilities for church
involvement. The Bibliography gives additional resources for further
study and suggestions for action.

## The Power to Be a Parent

What follows is a fantasy. You are asked to try to put yourself in
someone else's shoes and feel the frustration, the agony, the anger,
and the pain often experienced by people who want children very
much and have not been able to conceive. While the leader reads
this fantasy aloud (slowly!), close your eyes and relax. Forget who
you are, and picture yourself . . .

You are now twenty-nine years old. You grew up in a large and
generally happy family, and you still remember how everybody pitched

in to help and how there was always somebody you could turn to in a crisis. Your own parents are the most important role models in your life. When you were little and "played house," you always wanted to be the "mommy" or the "daddy." You looked forward to marriage and to starting your own family. And you've always wanted a large family like the one you knew as a child.

When you got married seven years ago, you both agreed to work for two years until you were on solid footing financially so that you could start your family. You scrimped and saved and planned and dreamed. You bought a small house and painted the extra bedroom, getting it ready as a nursery.

Then you began trying to conceive. A few months went by, and you were surprised. You expected pregnancy to occur as soon as you stopped using contraception! Most of your friends were having babies, and you began to feel left out. After a while you began charting the months and becoming depressed when with each month it became clear that conception had not occurred. You began to focus on two times of the month: the time of ovulation when you would "try again," and the time when you hoped menstruation would not begin, signalling another failure.

By now, your home is full of charts and graphs, the thermometer sits by the bed, and sex has become laden with fears and anxieties. Doctors are still testing you and trying different hormones to see if anything will help. While you struggle with this situation, the following scenarios begin to happen. See how you feel when . . .

Your friends talk about completing their families, debating whether to stop at two or three children. How do you feel?

Your mother asks when you are going to give her a grandchild. What is your response?

Your younger brother has his fourth—and unexpected—child. He complains about diapers and being kept awake at night. What do you want to say to him?

You join a church. People ask you, "Do you have children?" "Where do you work?" Most of the adults your age attend the "family service" with their children. How do you feel there?

Finally, after a long series of medical tests, you learn that you will never be able to have children "of your own." How do you feel? How do you feel towards your spouse? What dreams do you have now for your future?

What would you like to say to your friends, to your mother, to your brother, and to all those well-intentioned church people?

## DISCUSSION QUESTIONS

1. Reflecting on the fantasy and your effort to identify with the people who could not bear children, what sorts of personal responses or congregational activities would be helpful to such a couple? What would be unhelpful—or even hurtful?

2. Consider your own experience and values, and the patterns of activity and expectations in your own church around marriage, families, and children:
   - what is the "norm" for families in terms of having children?
   - what family situations stand out enough to elicit comment? (e.g., "I wonder why they don't have any children?")
   - do your activities and structures as a church affirm and open up to families without children or do they exclude these families?

3. Brainstorm a list of possible activities and programs your congregation might undertake in response to couples facing these difficult situations. Make three lists:

   SUPPORT     EDUCATION     ADVOCACY

   Try to generate some specific ideas that your group or congregation might implement—e.g., a "singles" day that looks at different meanings of "family," a study group on tax laws for couples with and without children, etc. How might you incorporate some concerns about new technologies for expanding the range of people who can become parents?

## Normality, Disability, and Handicap

The deep desire to have children is usually for children conforming to our image of a "normal" baby. Efforts to employ new genetic technologies to avert the birth of a genetically "defective" child reflect our desires and preferences. If we could, we would probably always have "normal" or "healthy" children, without debilitating physical, medical, or mental handicaps.

But we lack the power to ensure that all of us will be healthy. Genetic diversity is important to the human race; but that same diversity means that there will always be a risk of bearing some children with disabilities. Through choice, and by accident, children who are mentally retarded or physically handicapped will continue to join our communities as infants and to mature in their own fashion in our midst.

This section invites us to consider our response to these people, and hence our response to parents struggling with questions of "quality control" in their offspring.

1. First, identify your own experiences with handicapped persons: We suggest that you do this individually and then share in small groups. Be as honest with yourself as you can.

   Do you remember the first time you saw somebody who "looked different" or was crippled? What were you told about how to respond?

   How often do you encounter handicapped people now? Where? How do you respond?

   Do you react differently to different sorts of handicaps? If your child had to have a physical or mental disability, which would you choose: blindness, deafness, mental retardation, or spasticity? What does your choice reflect about what you value?

   Do you have any friends who are handicapped? What have you learned from watching how they handle their handicap? Can you talk with them about God and God's role in their lives?

2. Make two lists, brainstorming as many possibilities as you can for each:

| Things we could do to close out handicapped persons, to deny their humanity. | Things we could do to reach out and include handicapped persons, to affirm their humanity. |
|---|---|

Now turn to the discussion questions following to assist further reflection.

### DISCUSSION QUESTIONS:

1. What do your own experiences with handicapped persons and your lists tell you about your attitudes and values? Can you identify values that you want to affirm and strengthen? Values and attitudes that you would like to change?

2. Try to imagine yourself as the parent of a child with a physical disability; a mental disability. What would you want most from the congregation? What would be most hurtful or helpful? Try the same process for a handicapped adult.

3. Our experience with handicapped persons presents a paradox. We support parents who seek genetic counseling or screening in order to avoid having handicapped children, and we share a sense of loss and sadness with the family who unexpectedly bears a child with a disability. Yet we also want to affirm and support handicapped persons in our midst. The paradox, put starkly, looks like this: We don't want handicapped persons, yet we want to affirm them. Can we do both at the same time? Reflect together on this paradox and on how you understand your own values and aspirations in light of all the material presented in this book.

4. What can your particular congregation do in the areas of support, education, and advocacy to respond to handicapped persons in your community? What do you know about genetic counseling and screening programs, and what might you do about new genetic technologies that offer "quality control" to parents? Brainstorm some ideas and try to propose specific steps for follow-up by the congregation.

These last discussions close our study process where it began—in the conviction that we are called to enact the loving bonds of community in response to the needs of parents as they face difficult decisions and, with the help of the community, attempt to live with these decisions for a lifetime. Our hope is that the discussions in these last two chapters will encouarge study groups to begin making decisions and taking concrete action.

# BIBLIOGRAPHY

## General Works in Bioethics

The following books provide a helpful introduction to principles in biomedical ethics and to various perspectives on issues such as abortion, genetic screening, the delivery of health care, etc.

Beauchamp, Tom L., and Childress, James F. *Principles of Biomedical Ethics*. New York: Oxford University Press, 1979.

Beauchamp, Tom L., and Walters, LeRoy. *Contemporary Issues in Bioethics*. 2d ed. Belmont, Ca.: Wadsworth Publishing Company, 1982.

Hunt, Robert, and Arras, John. *Ethical Issues in Modern Medicine*. Palo Alto, Ca.: Mayfield Publishing Company, 1977.

Reich, Warren T., ed. *Encyclopedia of Bioethics*. 4 vols. New York: Macmillan, 1978.

## Bioethics: Particular Perspectives

The following volumes represent particular perspectives on bioethical issues, from an explicit Christian perspective (Ramsey, Smith and Vaux), more popularized views (Restak), and a concern for feminist and women's perspectives (Holmes, et al).

Curran, Charles. *Issues in Sexual and Medical Ethics*. South Bend, Ind.: University of Notre Dame Press, 1978.

Häring, Bernard. *The Ethics of Manipulation.* New York: Seabury Press, 1975.

Holmes, Helen B.; Hoskins, Betty B.; and Gross, Michael. *The Custom-Made Child? Women-Centered Perspectives.* Clifton, N.J.: The Humana Press, 1981.

Ramsey, Paul. *The Patient as Person.* New Haven, Ct.: Yale University Press, 1970.

Restak, Richard M. *Premeditated Man: Bioethics and the Control of Future Human Life.* New York: Penguin Books, 1973.

Smith, Harmon L. *Ethics and the New Medicine.* Nashville: Abingdon Press, 1970.

Vaux, Kenneth. *Biomedical Ethics: Morality for the New Medicine.* New York: Harper and Row, 1974.

## Genetics

The following books and materials represent discussion of ethical issues in the development of new technologies in genetics. They include specific treatments of ethical issues in genetic screening, prenatal diagnosis and selective abortion, and the development of technologies such as artificial insemination and in vitro fertilization.

Birch, Charles, and Abrecht, Paul. *Genetics and the Quality of Life.* Elmsford, N.Y.: Pergamon Press, 1975. A publication of the World Council of Churches.

Bergsma, Daniel, ed. *Ethical, Social and Legal Dimensions of Screening for Human Genetic Disease.* New York: Symposia Specialists, 1974. A publication of the National Foundation/March of Dimes.

Fletcher, John C. *Coping with Genetic Disorders.* San Francisco: Harper and Row, 1982.

Fletcher, Joseph. *The Ethics of Genetic Control: Ending Reproductive Roulette.* Garden City, N.Y.: Anchor Books, 1974.

Hamilton, Michael, ed. *The New Genetics and the Future of Man.* Grand Rapids, Mi: Eerdmans, 1972.

Harris, Maureen, ed. *Early Diagnosis of Human Genetic Defects: Scientific and Ethical Considerations.* U.S. Government Printing Office, HEW Publication number (NIH) 72–25.

Hilton, Bruce, et al, eds. *Ethical Issues in Human Genetics: Genetic Counseling and the Use of Genetic Knowledge.* New York: Plenum Press, 1973.

Judson, Horace Freeland. *The Eighth Day of Creation.* New York: Simon and Schuster, 1979.

Lappé, Marc. *Genetic Politics.* New York: Simon and Schuster, 1975.

Lipkin, Mack Jr., and Rowley, Peter T. *Genetic Responsibility: On Choosing Our Children's Genes.* New York: Plenum Press, 1974.

Lynn, Barry. *Genetic Manipulation.* Futures Papers No. 4. New York: Office for Church in Society, United Church of Christ, 1977.

National Council of Churches. *Human Life and the New Genetics.* New York: National Council of Churches, 1980.

Ramsey, Paul. *Fabricated Man: The Ethics of Genetic Control.* New Haven, Ct.: Yale University Press, 1970.

Roslansky, John D., ed. *Genetics and the Future of Man.* New York: Appleton-Century-Crofts, 1966.

Wade, Nicholas. *The Ultimate Experiment.* 2d ed. New York: Walker & Co., 1979.

Wolstenholme, Gordon, ed. *Man and His Future.* Boston: Little, Brown, 1963.

## Parenthood

There are not many books dealing explicitly with questions of genetics and parenthood. The following volumes give philosophical perspectives on parenthood, discuss problems in raising children with genetic anomolies, and provide an overview of issues in Christian tradition around the link between sexuality and parenthood.

Allen, David F., and Victoria S. *Ethical Issues in Mental Retardation.* Nashville: Abingdon Press, 1979.

Kilman, Gilbert W., and Rosenfeld, Albert. *Responsible Parenthood: The Child's Psyche Through the Six-Year Pregnancy.* New York: Holt, Rinehart and Winston, 1980.

McGinnis, Kathleen and James. *Parenting for Peace and Justice.* Maryknoll, N.Y.: Orbis, 1981.

O'Neill, Onora, and Ruddick, William. *Having Children: Philosophical and Legal Reflections on Parenthood.* New York: Oxford University Press, 1979.

Ross, Bette M. *Our Special Child: A Guide to Successful Parenting.* New York: Walker and Company, 1981.

United Church of Christ. *Human Sexuality: A Preliminary Study.* New York: United Church Press, 1977.

U.S. Department of Health and Human Services. *Learning Together: A Guide for Families with Genetic Disorders.* U.S. Government Printing Office, DHHS Publication number (HSA) 80–5131.

## Christian Ethics

The following volumes are only some of the many approaches to Christian ethics that might help a church group formulate its own position on crucial issues in genetics and parenthood. They represent different approaches to the field of ethics and suggest several alternative methodologies. They also include reflections on the use of scripture, the meaning of the virtues in human life, the place of rules and of the situation in ethical decision making, and the role of self in making ethical decisions.

Fletcher, Joseph. *Moral Responsibility: Situation Ethics at Work.* Philadelphia: The Westminster Press, 1967.

Gustafson, James M. *Christian Ethics and the Community.* New York: The Pilgrim Press, 1971.

Gustafson, James M. *Ethics from a Theocentric Perspective.* Chicago: University of Chicago Press. Vol. 1: Theology and Ethics, 1981. Vol. 2: Ethics and Theology, expected in 1983 or 1984.

Gustafson, James M. *Theology and Christian Ethics*. New York: The Pilgrim Press, 1974.

Hauerwas, Stanley. *A Community of Character*. Notre Dame, In.: University of Notre Dame Press, 1981.

Hauerwas, Stanley. *Truthfulness and Tragedy*. Notre Dame, In.: University of Notre Dame Press, 1977.

Marty, Martin E., and Vaux, Kenneth, eds. *Health/Medicine and the Faith Traditions: An Inquiry into Religion and Medicine*. Philadelphia: Fortress Press, 1982.

Niebuhr, H. Richard. *The Responsible Self: An Essay in Christian Moral Philosophy*. New York: Harper and Row, 1963.

Ramsey, Paul. *Deeds and Rules in Christian Ethics*. New York: Charles Scribner's Sons, 1967.

Shinn, Roger L. *Forced Options: Social Decisions for the 21st Century*. San Francisco: Harper and Row, 1982.

Wogaman, J. Philip. *A Christian Method of Moral Judgment*. Philadelphia: The Westminster Press, 1976.

## Other Resources

In addition to the published material listed above, there are a number of organizations providing specialized resources. Local hospitals and children's hospitals usually know about organizations that help parents whose children have Down's syndrome, cystic fibrosis, sickle-cell anemia, and other disorders.

The Hastings Center, 360 Broadway, Hastings-on-Hudson, New York, N.Y. 10706, publishes a monthly *Report* dealing with issues in bioethics, and also provides other services.

The National Foundation/March of Dimes, 1275 Mamaroneck Drive, White Plains, N.Y. 10605, is well known for its work toward the prevention of birth defects.

Groups such as the National Association of Retarded Citizens and the Black Child Development Institute provide resources for families to deal with mental retardation and other special problems.

Most major medical centers can provide information about new technologies such as in vitro fertilization or genetic screening programs. Individuals, parents, and church groups should check their local organizations for resources on genetics, ethics, and parenthood.